Anatomy & Physiology

Made Easy

An Illustrated Study Guide for Students To Easily Learn Anatomy and Physiology

Published by NEDU LLC

Written by the creators at

Disclaimer:

Although the author and publisher have made every effort to ensure that the information in this book was correct at press time, the author and publisher do not assume and hereby disclaim any liability to any party for any loss, damage, or disruption caused by errors or omissions, whether such errors or omissions result from negligence, accident, or any other cause.

This book is not intended as a substitute for the medical advice of physicians. The reader should regularly consult a physician in matters relating to their health, and particularly with respect to any symptoms that may require diagnosis or medical attention.

NCLEX®, NCLEX®-RN, and NCLEX®-PN are registered trademarks of the National Council of State Boards of Nursing, Inc. They hold no affiliation with this product.

Some images within this book are either royalty-free images, used under license from their respective copyright holders, or images that are in the public domain.

Illustrations created by our wonderful team. The content was written by the creators of NurseEdu.com. Published by NEDU LLC.

For bulk orders in paperback, please contact: Support@NurseEdu.com

© Copyright 2021, NurseEdu.com. All rights reserved.

ISBN: 978-1-952914-16-4

FREE BONUS

FREE Download – Just Visit:

NurseEdu.com/bonus

Table of Contents

Chapter One: The Human Body

So, you've decided to study the human body and you want to know where to begin. You're in the right place! In this book, you will learn anatomy and physiology, complete with stories and analogies to make learning easy. Once you've read this book, you'll look back and realize that you've learned this part of medical science in a fun and easy way. Each section includes illustrations to provide another way for you to learn. Whether you're interested in learning about the human body or studying for your A&P exams, this book contains everything you'll need to know to succeed. This book also includes over 300 illustrations to truly make this an illustrated study guide.

In this beginning section, you'll first learn the basics in order to identify the different parts of the body. There are different planes, regions, directional terms, and cavities to learn about. In the chapters that follow, you'll study new concepts about the cells that make up our bodies, even though these things are too small to be seen with the naked eye.

Why study cells in anatomy? One thing you'll soon figure out is that molecules make cell structures, cell structures make cells, cells make tissues, and tissues come together to form the organs of the body. There is no cell or organ of the body that works just by itself. They all communicate with one another in ways that make your body a dynamic machine.

It's all quite amazing if you think about it. The various body parts might seem complicated, but as a whole it's very simple. The body is designed so well that from head to toe, the whole thing works together on a 24/7 basis to keep processes running smoothly from birth until the end of life. So, are you ready to explore what the human body looks like on the inside and how it works? If so, just read on.

Anatomy is the Structure of the Body

From the title of this book, you can see there are two parts: anatomy and physiology. What do these really mean? Anatomy is the "what" of everything. *What is this part called? What are the connections between part A and part B in the body? What are the different parts of the respiratory system, the gastrointestinal system, and the skeletal system?* It is the study of the body's structures.

The first thing you'll notice when you look at yourself in the mirror is that you have a head and a neck, a torso with an upper and lower part, and four limbs (your arms and legs). How is it that you can stand up and actually see yourself in the mirror? That's because, on the inside, you have a remarkable skeleton made up of your bones. All the major bones that keep you upright are connected by joints and cartilage.

Of course, you're not just your bones. You've got meat on your bones, for example. That's your muscles and fat, all neatly packaged inside your skin. Did you know your skin is the largest organ of your body? Inside of your skeleton, and protected by it in some cases, are your *internal* organs, which are sometimes called *viscera*. Some people think the viscera is just your gut, but in anatomy viscera refers to every internal organ, including your lungs and heart.

Another main reason you're able to stand up is because you have a spinal cord. It's housed in a stack of disc-shaped bones called vertebrae that also protect it. You really need your spinal cord, because it's the connecting highway between your brain (which is where you think about what you want to do, and perceive sensory information from the environment) and the rest of your body. Your muscles are what make you move, but without the nerves from your spinal cord they won't do anything. That's the issue paraplegics and quadriplegics have, because their spinal cord is damaged. The spinal cord houses all the nerves that give you your senses, and tell each muscle where you want to move and by how much.

The study of anatomy is basically the study of names. It's important to learn the names of bones, muscles, nerves, blood vessels, and other parts of the body. You'll quickly see that there's a method to the madness. Once you learn the tricks to how things are named, it will make a lot more sense. You'll see why, for example, the superior mesenteric artery is different from the inferior mesenteric artery, as well as why they're named the way they are. Once you figure these things out, a lot more about anatomy and the naming of things will make sense.

Fun Factoid: Did you ever wonder who first figured out what was inside the human body? Well, the unofficial "father of anatomy" is Herophilus, who lived around 300 BCE in Alexandria, Egypt. Yes, he did study the human body to a great degree, but his work was really controversial because he learned on living people who were cut open to study their insides. This was taboo at the time and probably not any fun for the "specimens," who likely weren't knocked out before they were dissected. Many years later, an Italian doctor named Andreas Vesalius published the first modern anatomy book in 1543, so he was given the official title of the "father of modern anatomy."

Physiology Makes Your Anatomy Do Things

The second aspect of this book is "physiology." This is related to anatomy but really isn't the same thing at all. If anatomy is the "what" of your body parts, then physiology is everything else. Your anatomy will always be a part of you, alive or dead, but physiology is what your body does only when it is a living thing.

Physiology can be biochemical. It can include how the different molecules get inside or outside the cells, and explain how each of the tiny molecular processes happen inside the organelles, or internal structures of your cell. Your body is a giant electrical system, mechanical system, and chemical system all at once.

Through your body's physiology, each and every cell and organ has chemical communication systems that talk to each other. As you'll see, some cells just talk to their neighbors, while others talk to cells practically all over your body. Did you know, for example, that your gut talks to your brain, giving it signals that affect your mood? Yes, all the parts of your body really do connect to one another, which is another way you're like a well-oiled machine when everything works the way it's supposed to.

Fun Factoid: Speaking of your gut, did you know your poop can talk too? The "good" bacteria in your poop sends signals that can affect your brain and immune system. If you have a severe gut infection that destroys your healthy bacteria, your doctor might recommend a "poop transplant," which means you have to take pills containing other people's fecal matter. The idea is that there are healthy bacteria in the capsules that repopulate your gut, so you'll have good bacteria after that. That's just gross, but it's effective!

One term you should get used to understanding when it comes to physiology is *homeostasis*. This basically means "stable," and is what the physiology of your body is trying to accomplish on a 24/7 basis.

Your body wants stability and balance all the time, so it has a huge and intricate system of checks and balances called *feedback systems* that do things like keep your blood sugar regulated, your temperature within a normal range, and your blood pressure just high enough to get your blood from your head to your toes. All of this stability, or homeostasis, happens on a cellular level as well as on a total-body level.

Directional Terms in Anatomy

Let's take a look at the different directional terms you'll need to memorize when you study human anatomy. These are the terms that tell you which things are higher than others on the body, closer or further away from the body's center, and everything else you'll need to know in order to figure out how the different anatomical parts are related to one another. These are the directional terms that will most help you in your study of anatomy:

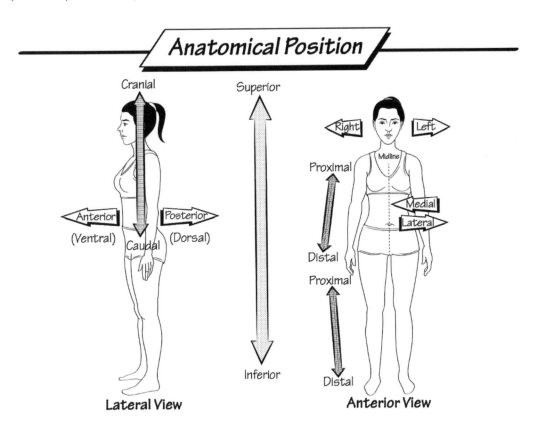

- **Inferior and superior:** *Inferior* refers to something that is below a particular point, while *superior* refers to something that is above a particular point. Your head is the most superior part of your body; your feet are the most inferior part of your body.
- **Anterior and posterior:** *Anterior* is another way of saying something is toward the front of the body. *Posterior* is another way of saying something is toward the rear of the body. Your stomach is anterior and your back is posterior.
- **Medial and lateral:** *Medial* means nearer to the center or midline of the body, while anything *lateral* is further away from that same center or midline. Draw a line down the center of your body, dividing it into right and left halves. That's your midline.
- **Distal and proximal:** *Proximal* is the word you will use when talking about something that is close to the trunk of the body, while *distal* is the word you will use when talking about

something that is farther away from the trunk of the body. Most of the time, you'll use these terms when you talk about your arms or legs. The hand, for example, is distal to the shoulder because it is further away from your trunk.

- **Superficial and deep:** Anything *superficial* is near the surface of the body. Anything *deep* is far from the surface of the body. You will learn about superficial veins that you can see because they are close to the body surface, and deep veins you cannot see because they are far from the surface of the body.
- **Intermediate:** This refers to anything that's between two reference points. For example, the heart is called *intermediate* when you are talking about its location, because it's between the two lungs.
- **Caudal:** This term is unusual because it can relate to animals and humans alike. *Caudal* means that something is lower than another part of the body, or towards the posterior part of the body, like a tail.
- **Cranial:** This is the opposite of caudal and means anything near the head. Your cranium, for example, is the medical term for your head cavity.
- **Visceral:** This is something you might see instead of the word *deep*. It means anything not visible on the surface of the body. Your viscera, for example, are your internal organs.
- **Palmar:** Refers to the palm side of the hand.
- **Plantar:** Refers to the bottom of the foot.
- **Dorsal:** This means something is on the back of a reference point. A nerve that's viewed as *dorsal* is toward the back. The word *dorsal* is sometimes interchangeable with *posterior*.
- **Ventral:** This is opposite of dorsal and means the front of a reference point. *Ventral* is sometimes interchangeable with *anterior*.

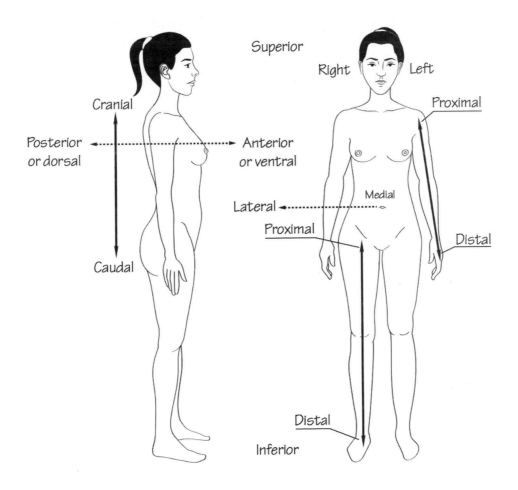

Body Planes

Knowing the different body planes is extremely helpful when you're trying to describe where something is located in the body. You might, for example, see the word *parasagittal* and wonder what it means. Basically, it means "near the sagittal line." If you don't know where that line is, you'll really be lost.

Body planes help divide the body into different sections that are more precise than saying "just to the left" or "down the middle." Remembering the planes ensures everyone is on the same page about the exact location. These are the three body planes you should remember:

- **Sagittal plane:** This is also called the *lateral plane*. It separates the body into left and right halves. The midsagittal plane is the midline of the body. A parasagittal plane is any line that is parallel to the midsagittal plane. The term *sagittal* means "arrow" in Latin, so just imagine shooting an arrow between your eyes to separate the left and right sides of your head.
- **Coronal plane:** This is also called the *frontal plane*. It divides the body into the front and back parts, which are also called the dorsal and ventral parts of the body. The term *coronal* means "crown" in Latin. Just imagine a tiara that divides the princess's head into the front half and back half.
- **Transverse plane:** This is also called the *axial plane*. It divides the body into the top half and the bottom half. It is the only one of the planes that is horizontal and parallel to the ground. This is

related to the term *longitudinal plane*, which is any plane that cuts the body into a top half and a bottom half. The term *transverse* means "across" in Latin, so imagine sawing across the middle of the body, cutting it into two pieces.

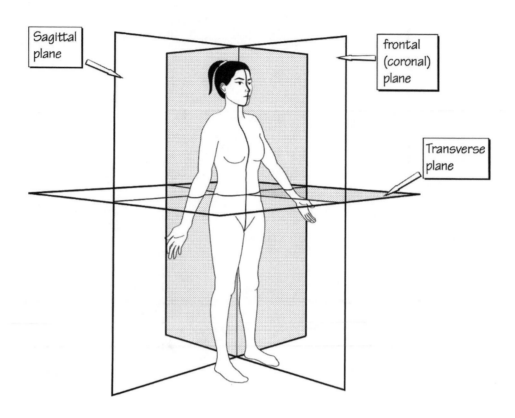

Body Cavities

Your body is divided into cavities, or spaces, that keep all your organs separated. Usually, they separate the body into spaces that have organs involved together doing similar things, but not always. For example, your abdominal cavity holds your abdominal viscera, like your stomach, small intestines and colon. It also houses your liver and gallbladder, which are important to your gastrointestinal system as well, even though food doesn't go through them directly.

Like knowing the body planes, studying the body cavities can help you realize we aren't just a bag of bones. Instead, we have real physical separations between our organs. The different body cavities are physically separated from one another by membranes, or sections of connective tissue. These keep your organs in place so they don't shift around that much when you move.

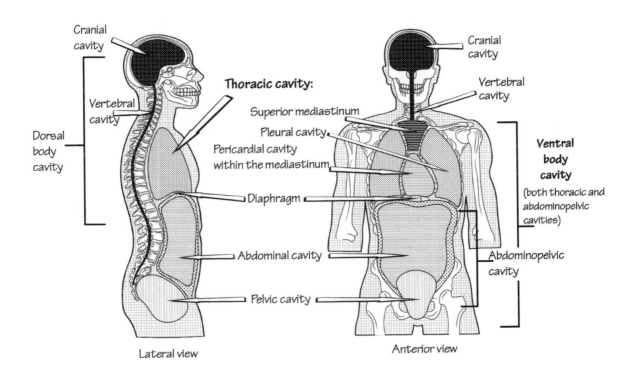

Take a look at this image of the cavities so you can see how your body is divided up. There is a smaller dorsal body cavity and a much larger ventral body cavity. Remember that dorsal refers to the back and ventral to the front. Each of these is itself further divided.

The dorsal body cavity is divided into the cranial cavity and the vertebral cavity. The cranial cavity is where your brain is located and the vertebral cavity is where your spinal cord is located. These are technically connected to one another without a membrane between them, and thus are divided for convenience only.

The ventral body cavity has three main parts. Your heart and lungs, plus the major vessels around the heart, are in the *thoracic* cavity. The *abdominopelvic* cavity is huge but is itself divided into the abdominal cavity and the pelvic cavity. The abdominal cavity is for your gastrointestinal tract, or GI tract, mainly the areas from your stomach to your anus. The pelvic cavity is below the abdominal cavity and is where the female reproductive organs and bladder are located.

The thoracic cavity is also divided into three parts. They are usually separated by membranes that keep the structures of the chest in the right place. The *superior mediastinum* is the upper part, containing a lot of arteries, veins and nerves. The *pleural* cavity is where your lungs are. The *mediastinum* is in the middle, where you'll find the heart.

In some cases, the separating structures between two nearby cavities is obvious. An example is the diaphragm. It is a huge, thick muscle that separates your abdominal and thoracic cavities. In other cases, there is just a thin membrane. The pericardium, for example, is a membrane that surrounds the heart and defines where you'll find the mediastinum. Between the abdominal cavity and the pelvic cavity, though, there is nothing. The idea of separate "cavities" is kind of arbitrary here, because there isn't really any physical separation between the two.

More Divisions to Consider: The Different Regions of the Body

There are two main regions of the body you should know about, and several divisions within these regions. Let's look at the two main regions first:

- **Axial region:** This refers to the middle of your body but not your arms or legs. It is divided into four separate sub-regions, which are the head, neck, chest and trunk regions.
- **Appendicular region:** This refers to just your arms and legs. You can divide this region up into the limbs (your arms and legs) and your appendages (your fingers and toes).

Fun Factoid: You can remember the appendicular region because it involves appendages. The definition of an appendage is confusing, though, because it means more than just your arms, legs, fingers and toes. It can be anything that sticks out from the body. If you were a snake, your fangs would be considered appendages; technically, your ears stick out and would be appendages, too, but these are not a part of your appendicular region because they are stuck to the side of your head. The term appendix is derived from the same root word, but it also is not a true appendage, in the sense that it isn't part of your appendicular region.

In actuality, it can get even messier than this. The anatomists who named all of the parts of the body were really big on narrowing things down to the tiniest of divisions. As a good exercise, let's look at how some of these main regions are divided up. Take the cranial region, or head area, for example. Anatomists have decided that there are ten different regions in this area alone! These include some you may or may not recognize:

- The *cranial region* is just your head minus the face, while the *facial region* is the face below the level of the ears.
- The forehead is the *frontal region*.
- Below that is the *orbital* or *ocular region*, which is the area around your eyes.
- The ears are the *otic* or *auricle region*.
- The area around your nose is the *nasal region*.
- The cheeks are the *buccal region*.
- The mouth itself is the *oral region*.
- The chin is called the *mental region*.
- The neck is the *cervical region*.

Fun Factoid: Why is your chin called the mental *region? Isn't* mental *related to your mind? In Latin, there are two related words:* mentum *and* mens. *The term* mentum *means "chin" and the term* mens *means "mind." Word researchers think they are related to the idea that, when you think about something, you sometimes rub your chin. So, rub your chin and ponder that connection!*

Let's try dividing up another major region so that you can see how this is supposed to work. Again, knowing a region's subdivisions will help you figure out which smaller area of a larger region is being discussed. For the posterior regions of the legs, there are six subdivisions:

- Your buttocks are your *gluteal region*.
- Your thigh is the *femoral region*.

- Your knee is the *popliteal region*.
- The back of your lower leg is the *sural region*.
- Your heel is called the *calcaneal region*.
- The sole of your foot is called the *plantar region*.

Fun Factoid: You can remember your gluteal region if you are an athlete trying to strengthen your glutes. The word is Greek, stemming from the word gloutos *in the Greek language, which means "buttock." Yeah, it's your butt.*

Again, knowing these different subdivisions or subregions is a way to narrow down a description of where something is located. You can for example say, "The patient has pain in his calcaneal region," instead of saying "his heel hurts." It just sounds better from a professional sense to use these anatomical terms, rather than the common names for these regions.

The Abdominal Quadrants

The abdomen is a special region that can be looked at in two ways. The first is just the four quadrants—right upper, right lower, left upper and left lower. These are helpful in medicine, because pain in different areas of the abdomen can mean different things.

If you include the pelvis along with the abdomen, there are nine different subregions. These subregions are named and can be seen in the figure shown. Let's go down the list:

- **Right and left hypochondriac regions**: These are the uppermost and most lateral regions. On the right is the liver and part of the upper kidney. On the left is part of the pancreas and the spleen.
  Fun Factoid: There is a reason why the word hypochondriac, *meaning you are super-anxious about your health, and the anatomical term* hypochondriac *are the same. This is because the ancient Greeks thought that this kind of anxiety was due to a problem with your spleen, which happens to be located in this area.*
- **Epigastric region:** This is the upper middle section where your stomach is located. Your abdominal aorta, part of the pancreas, the duodenum, the inferior vena cava, and part of the liver can be found here.
- **Right and left lumbar regions:** These are the middle and lateral regions. Most of the kidneys on either side, and parts of the colon on both sides, are seen here.
- **Umbilical region:** This is the area around your belly button. Most of your small intestine, or small bowel, are here, although some of the mid-colon can be seen here too.
  Fun Factoid: The Greeks had a word called omphalos, *which is where the word* umbilical *comes from. In Greek, the whole thing refers to the* navel, *which is your belly button.*
- **Right and left inguinal, or iliac, regions:** These are the lowest and most lateral of the nine subregions. They have parts of the colon on both sides, some of the small bowel, and sometimes the ovaries in women.
- **The hypogastric region:** This is also called the pubic region. You will find the sigmoid colon, the bladder, the female ovaries, and the female uterus here.
  Don't Be Confused: If you know your medical terms, you know that hypogastric means "below the stomach." It comes from the Greek word hypogastrion, *which has the same meaning. What they don't say is that it is way below the stomach—so far below that it's really in the pelvic area.*

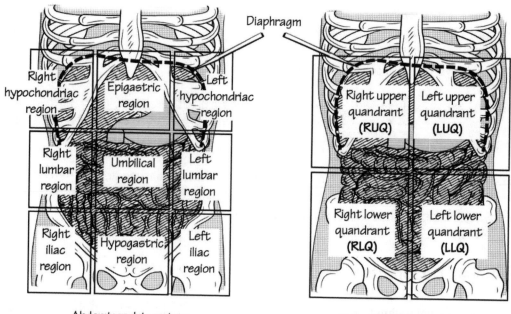

Diaphragm

Right hypochondriac region | Epigastric region | Left hypochondriac region

Right lumbar region | Umbilical region | Left lumbar region

Right iliac region | Hypogastric region | Left iliac region

Abdominopelvic regions

Right upper quandrant (RUQ) | Left upper quandrant (LUQ)

Right lower quandrant (RLQ) | Left lower quandrant (LLQ)

Abdominopelvic quandrants

As we talk about the organs of the abdomen and pelvis later, you should be able to judge where they are located in the different subregions. You should know that the abdominal area is also divided into the parts protected by the peritoneum, which is a membrane that covers most of the abdominal contents, and the parts behind the peritoneum in what's called the retroperitoneal space. Your kidneys, for example, are in this retroperitoneal space. In the pelvis, this peritoneal membrane drapes over the uterus and bladder, but they are technically located outside the peritoneum itself.

Don't Be Confused: *You might think that the peritoneum just encloses the abdominal contents like a sac, but this isn't true. Instead, it's more like a curtain draped over the abdomen. This explains why you'll see parts of the bladder with peritoneum on it, where it has draped over this organ, and parts that don't have peritoneum, like the backside.*

Inside the abdomen and within the peritoneum, things don't just jostle around in there randomly. There are membranes and ligaments that hold everything in place. One example is the lesser sac and greater sac of the stomach area. These are also called the greater and lesser *omenta*.

Fun Factoid: *The omentum is not the same as your peritoneum, but is actually made out of the peritoneum itself. Another term for omentum is* caul*, which is Middle French for* cale*. This means "cap" and is the same as the caul covering some babies' heads if they are born with their birth membranes still capping their head. The caul and omenta are made from the same basic connective tissue material.*

The picture below shows the omenta. Do you see how they hold the stomach in place? Color the illustration below and highlight the Greater and Lesser Omentum.

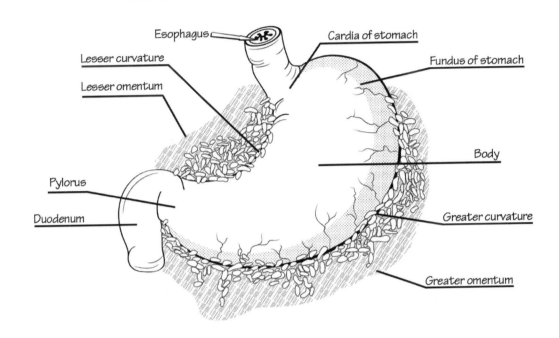

Anatomy of Stomach

Esophagus

Cardia of stomach

Fundus of stomach

Lesser curvature

Lesser omentum

Body

Pylorus

Greater curvature

Duodenum

Greater omentum

There are other ligaments you will learn about that are designed to hold the different abdominal and pelvic structures in their proper place.

To Sum Things Up

Now that you have navigated the basics of the human body when it comes to its directions, planes and regions, you can start your study of anatomy and physiology with a better idea of how the body is divided. You will see how descriptions of anatomical terms use these basic ways of describing the human body, especially using directional terms and body regions or subregions. Memorize these now and you will save yourself a lot of trouble learning the different terms later.

Chapter Two: Recognizing Anatomical Terms

In this chapter, we'll introduce some key medical terminology, and the root words behind the anatomical terms in this book.

Medical terminology is like learning your ABCs in school, when what you really wanted was to put them together to make real words. If your teacher had handed you a copy of *War and Peace* the day you started kindergarten, though, it would have seemed like a lot of gibberish and you wouldn't have learned much. Instead, you started with the basics and worked your way up from there.

This is exactly what you have to do with medical terminology. These are parts of words that mainly came from the ancient Greeks and Romans, who were fond of building words based on the meaning of each part. Exactly why doctors and people who study anatomy still use these terms today is a mystery. It's like a secret language that keeps the everyday person wondering why we use the word *cardiac* when we're talking about the heart.

Rather than dwelling on why this secret language hasn't evolved over time, just think of it as belonging to a special club of those who understand that the word *hypoglossal* isn't a swear word, but actually means something that makes a whole lot of sense when you understand its parts. (Spoiler alert: It means "under your tongue.")

There are hundreds of different anatomical terms you can learn and memorize. In this section, we'll go over some of the more important ones, according to the area of the body they belong to. Get out some blank flashcards as you read these next few paragraphs. On one side, write the anatomical root word you want to memorize. On the other, put its English meaning. For example, one side of the card could say *-ology*. The flipside of the card would then say "the study of."

Once you have a nice big pile of flashcards with medical terms on them, play a little game of medical terminology solitaire with yourself. Look at the root word and guess its meaning. If you get it right, put it to one side. If you get it wrong, put it on the other. When you are done, pick up just the pile you got wrong and start over. Do this until every card is in the "correct answer" pile. Then you can congratulate yourself on winning the game.

Repeat this exercise every so often, until you get almost all of the root words right the first time. Eventually you won't need any flashcards, and the secret language of medical terminology will seem effortless. You'll be speaking "doctor language" fluently.

As you study these terms, remember that these are mostly just parts of words that together make a whole medical term. Like words in English, medical terms often have a prefix, root word and suffix. The prefix is at the beginning. It can be a color term, a directional term, or anything else that modifies the root word, or main meaning. These are terms that translate into major body parts, such as those that refer to the heart, lungs, stomach or blood. The suffix is at the end of the word. It further directs you to what the meaning of the word is supposed to be.

After you study this chapter, you'll be able to look at the word *hematology*, for example, and draw a line between the root word and suffix like this: hemat/ology. You'll know that *hemat-* means blood and that

-ology means the study of something, making *hematology* the study of the blood and blood systems of your body.

Suffixes are more important than they might seem. If you mix up the medical terms *jejunostomy* and *jejunectomy* as a surgeon, you might remove part of the person's small bowel (the jejunum) when what you were supposed to do was just put a tube into it. Oops. Don't underestimate the suffix just because it is tacked onto the end of the root word.

Just as in grade school, when you separated long words into syllables and parts, you will do the same thing with medical terms. This will help you a great deal when you come into contact with an unfamiliar word. Identify the prefix, root word and suffix, and then make an educated guess at what it all means when put together. About 90 percent of the time, you'll get really close to the correct meaning.

As a word of caution, try not to get too hung up on exactness in medical terminology. For example, the medical terms *leuk-* and *leuko-* both mean *white*. The extra "o" at the end of *leuko-* just means that the word looks or sounds better with the vowel in it. There are rules about when to add or not to add a vowel, which really aren't very important as long as you recognize that these liberties with vowels are sometimes taken randomly in medical terminology.

EXTRA CREDIT: At the end of this chapter, we'll have a one-word quiz. You'll be presented with the medical mouthful of a term known as endoscopic retrograde cholangiopancreatography, *which is a relatively common gastrointestinal procedure. If you can figure out what it means, you can feel confident that you paid attention during this chapter. You can then consider yourself to be a medical terminology pro.*

Some Basic Prefixes

Prefixes can mean a lot of things. Most prefixes are what you could call directional. They tell you where to start your thought process when it comes to figuring out the rest of the word. Others are color-related prefixes. These too can give you clues as to what the entire word means. Still others are called number prefixes because they stand for a certain number you should pay attention to. Let's look at a few of these prefixes.

Prefixes for Direction

These help you decide where you are in relation to the root word. Are you to the left or right? Are you talking about something above or below it? These are all answered with directional prefixes.

- **Peri-:** This prefix means "around" or "surrounding." You use it when you say "perimeter," which is the length you travel around something, like a fence surrounding a yard.
- **Circum-:** This also means "around." The circumference of a circle is the length you'll travel when going around the entire circle.
- **Levo-:** This means "left."
- **Sinistr-:** This also means "left." Ancients used the term *sinister* as relating to the left side and something bad or evil, because left-handedness wasn't considered a virtue.
- **Dextro-:** This means "right."

- **Meso-:** This means "middle." The mesoderm layer in embryology is the middle layer of tissue, for example.
- **Medi-:** This also means "middle." Remember that medial is something in the middle of the body or toward the middle of the body.
- **Ab-:** This means "away from." If something is "abnormal," it means it is away from being normal.
- **Ad-:** This means "toward" or "near." The term *adverb* means a word that is near to a verb.
- **Dia-, per-, or trans-:** These all mean "through." The term *percutaneous* means to pass through the *cutaneous* area, which is the skin. The word *transatlantic* means to travel through the Atlantic Ocean.
- **Retro-:** This is a term that means "backward," "before," or "previous." If you use the term *retroverted* to describe a woman's uterus, it means the woman's uterus is tipped backward.

Prefixes for Position

These are helpful prefixes that are similar to directional prefixes, except that they are more static and don't imply any type of movement. They tell you where to look in relation to the root word in ways that are very similar to directional prefixes. Some positional prefixes mean that they are positional with respect to time rather than to physical location. Here are a few:

- **Ec- or ecto-:** This means "outside" or out of something. The word *eccentric* means outside of center, or someone who's a little "off center" in their thoughts or behavior.
- **Ex- or Exo-:** This also means "outside" or "away from." It is used in words like *exoskeleton*, which is the outside skeleton of insects.
- **Endo-:** This means "within" or "in." If you use the word *endometrium* in anatomy, you are talking about the inside layer of the uterus, because *metrium* refers to the uterus.
- **Epi-:** This means "upon" or sometimes "outer." The *epidural space* in the spine is outside the dura layer around the spinal cord.
- **Tele- or telo-:** This means "end." The word *telegraph* means to send a message from one end to the other. A *telomere* is a segment of DNA at the end of the chromosome.
- **Ante-:** This means "before" and can refer to time or physical position. The *antenatal* period in pregnancy is "before" the *natal* time, which is the time of birth itself.
- **Pre- or pro-:** This means "in front of" or "before." It is used in terms like *prodrome*, which is the time before you actually get sick. The *prenatal* period is also the same as the *antenatal* period, or the time before birth.
- **Post-:** This means "after" or "behind." It could also refer to time or position. Your posterior is also your behind, or backside.

Prefixes for Negating

These are a few prefixes that will come in handy when it comes to deciding if you are referring to the root word or perhaps the opposite, or negative, of the root word. For example, if you say someone is *moral*, this means something completely different than if you say they are *amoral*. Amoral means not moral at all, because the prefix "a" means "no" or "not." Let's look at some negative prefixes.

- **A- or an-:** This mean "not" or "without" and is a classic negative prefix. The medical word *anuria* means "no urination."
- **Anti- or contra-:** These mean "against." The word *contradict* means to oppose or be against someone in what you say, while the word *antisocial* means to be against social conventions or social norms.
- **De-:** This means down or without. The term *deescalate* means to tone down a confrontation.
- **Dis-:** This means the absence or removal of something. It can also mean to separate something. If you *dislocate* you shoulder, you separate your arm at the shoulder joint.
- **In- or Im-:** These both mean "not" but can also mean "into." If something is *impossible*, you are saying that it is "not possible."

Prefixes for Numbering

There are a lot of prefixes that refer to the number of something. It can be a specific number, or convey a general idea of a lot or a few. Some of these prefixes include:

- **Primi-:** This means "first." A primary election in politics refers to the first election rather than the final election, and primitive man is the first or "early" man.
- **Mono- or uni-:** This refers to "one" of something. A monocular lens is a device that just has a single lens, for example.
- **Bi- or di-:** This prefix means "two" or "twice." Your *bicycle* has two wheels (hopefully).
- **Tri-:** This means "three." A *tricycle* is when your bike has three wheels instead of two.
- **Diplo-:** This is a prefix used to describe "double" of something. Your basic *diploid* cell has two copies of each chromosome.
- **Tetra- or quadr-:** Both of these are used to describe four of something. Doctors might use the term *tetraplegic* and *quadriplegic* interchangeably, to refer to a person who is paralyzed in all four of their limbs.
- **Semi-** or **hemi-:** These both can mean "half," but are slightly different from one another. *Hemi-* often refers to one side of something, like the southern hemisphere or the southern half of the world. *Semi-* is somewhat different because it can also mean "partial," so the term *semiconscious* means you are only partially conscious.
- **Multi-:** This means "many." If you speak several languages, you are considered *multilingual*.
- **Poly-:** This also means "many" but could mean "much" in some situations. In geometry, for example, a *polygon* is a shape that has many sides to it.

Prefixes That Identify Colors

While it would be nice to just say "red cell" or "white cell," the people who made up terms used in anatomy and physiology just didn't think that simply. They made use of Greek and Latin terms to describe the color of things. This just gives you a few more things to memorize. Take these prefixes used to describe color as examples:

- **Chromo-:** This is the basic medical prefix for "color" in general. *Chromatography* in biochemistry is a technique that divides things according to their color.
- **Chlor-:** This is a prefix meaning "green." The *chlorophyll* inside plant cells is what make plants have their green color.

- **Cyan-:** This is a prefix meaning "blue." When you are *cyanotic*, it means you are turning blue from lack of oxygen.
- **Erythro-:** This is the prefix for "red." Instead of *red cell*, you might say *erythrocyte*, which actually does mean "red cell."
- **Leuko-:** This is the prefix for "white" and is used to describe a white cell, by calling it a *leukocyte* instead.
- **Melan-:** This is the prefix for "black." The *melanocytes* in your skin make the dark pigment that might give you a nice tan in the sunshine.
- **Xanth-:** This means the color "yellow." A strange yellowish-colored skin lesion called a *xanthelasma* is caused by the slight yellow color of cholesterol inside it.

Prefixes That Describe the Degree of Something

These are prefixes that help to describe if there is too much or too little of something. There are a few of these you should know about:

- **Hyper-:** This is a prefix meaning "over," "abnormally high," "excessive" or "increased." If you are *hyperactive*, for example, it means you are overly active in some way.
- **Hypo-:** This is a prefix meaning "under" or "below." It is a positional prefix if it means below. The hypoglossal nerve, for example, is below the tongue, which perfectly describes its position anatomically.
- **Oligo-:** This means "just a few" or refers to a scanty amount of something. If you are *oliguric*, in medical terms, it means you aren't peeing enough.
- **Pan-:** This is a prefix meaning "all." If you have *pancytopenia*, this means all your blood cells are too few in number, which is a bad thing for you.
- **Super or supra-:** These are terms that indicate being above something positionally, or that something is in excess. The word *superior* usually means something that is above another thing in anatomy.

Prefixes That Are Used for Size or Comparison

There are quite a few prefixes used in anatomy and physiology that help to compare things, in terms of their size or some other way. Let's look at these terms.

- **Eu-:** This is a prefix for "normal," but it can also mean "good," "true" or "easy." *Euthyroid* means that your thyroid gland is completely normal in terms of how it's functioning.
- **Hetero-:** This can mean "unequal" or "different" but can also mean "other." If you are *heterosexual*, you are mainly attracted to the other gender.
- **Homo- or homeo-:** Both of these mean "same" or "unchanging." *Homosexuality*, for example, refers to a preference for the same gender.
- **Iso-:** This means "the same" or "equal." An *isotope* in chemistry refers to atoms that have different atomic weights but are actually the same chemical type.
- **Macro- or mega-:** These terms both mean "large." A *macrocytic* cell, for example, is a very large cell in your bloodstream.
- **Megalo-:** This is related to the previous terms but tends to mean "abnormally large." A *megalomaniac*, for example, has an overly large or inflated sense of self.

- **Micro-:** This is a prefix that means "small," or technically one-millionth of something. If you have *microcytic anemia*, your red blood cells are too small.
- **Neo-:** This stands for "new." In medicine, a *neoplasm*, or cancer, is called by this term to reference the fact that this is new tissue.
- **Normo-:** This is a prefix meaning "normal." If you have *normocytic anemia*, you have anemia with red blood cells that are of a normal size.
- **Ortho-:** This is a prefix meaning "straight," "upright" or "correct." Your *orthopedist*, for example, helps make your bones straight again if they are broken.
- **Poikilo-:** This is a tough one not used in everyday language, that means "irregular" or "varied." If you see *poikilocytosis* on a microscope slide, it means you are seeing cells that are of different sizes and shapes.
- **Pseudo-:** This is the prefix for "false." Some single-celled organisms have *pseudopods* used for locomotion, which means they have false feet.
- **Re-:** This is a prefix meaning "again" or "back." *Revision* means to do something over again.

Some Miscellaneous Prefixes

There are a couple of miscellaneous prefixes used fairly often in medicine to describe certain conditions. There are a few of these to remember:

- **Ambi-:** This means "both" and is used in the term *ambidextrous*, which means that you equally use your left and right hands for doing things.
- **Dys-:** This is a prefix meaning "bad," "difficult" or "painful." This is a term that can be used for many things. If you have *dysmenorrhea*, you have painful periods.
- **Mal-:** This a prefix that means "bad" or "poor." If you are *malnourished*, your nutritional level is considered to be poor.

Some Suffixes You Should Know

Suffixes occur at the end of nearly every term in anatomy, physiology and medicine. They will help you define the root word in similar ways as prefixes. As you will see, prefixes and suffixes by themselves are sort of meaningless without a root word, but instead help to provide important context when looking at the root word itself.

Here are a few suffixes that might help you in your study of anatomy and physiology:

- **-algia:** This is a suffix that means "pain." You would use the term *arthralgia* if you're describing pain in the joints.
- **-emia:** This is a suffix meaning "blood." It is used in many blood-related terms like *anemia*, which literally means "no blood," and *hyperglycemia*, which means there is too much glucose in the blood.
- **-itis:** This is a common suffix term that means "inflammation." Just about any area of the body can be inflamed, so it's used in terms like *arthritis* and *dermatitis* to mean inflammation of the joints and skin, respectively.
- **-lysis:** This is a term that means to break down or destroy. The medical term *hemolysis* means the breakdown of the blood.

- **-oid:** This is a nice catchall term that means "like." You can describe something as *ovoid* if it is shaped like an oval in some way.
- **-opathy:** This is another common term used to describe any type of disease. The terms *neuropathy* and *enteropathy* respectively mean diseases of the nerves or gastrointestinal system.
- **-pnea:** This is a term that simply means "breathing" or "breath." It is used in terms like *hypopnea* or *apnea* to respectively describe slow breathing or lack of breathing.
- **-centesis:** This is a suffix that refers to puncturing a cavity in order to remove fluid. In an *arthrocentesis*, the doctor will puncture the joint cavity in order to remove fluid from it.
- **-ectomy:** This refers to the surgical removal of something. In an *appendectomy*, the surgeon will remove the appendix.
- **-ostomy:** This means to make a permanent opening into a body space. In a *tracheostomy*, for example, the surgeon will put a permanent hole into the person's trachea.
- **-otomy:** This means cutting into or incising something. In an *osteotomy*, the surgeon will cut into the bone itself.
- **-orrhaphy:** This is a suffix meaning to suture or surgically repair something. A *herniorrhaphy* is a surgical repair of a hernia.
- **-opexy:** This is an uncommon suffix meaning to surgically fix something into place. In a *nephropexy* procedure, the surgeon will fix the kidneys into a more solid position.
- **-plasty or -oplasty:** Both of these are terms that mean to surgically repair something. An *abdominoplasty* is commonly known as a "tummy tuck," or a *rhinoplasty* a "nose job."
- **-otripsy:** This is a suffix used for crushing or destroying something. One common term using this suffix is a *lithotripsy*, which involves breaking up kidney stones.
- **-ology:** This is a common suffix meaning "the study of." It's used everywhere and not just in medicine. The word *entomology*, for example, is the study of bugs or insects.

Recognizing Root Words

Root words are the meat of medical terminology, and tell you what body part you're dealing with. Some of these will be very familiar to you, while others will seem like a different language. In this section, we will divide the body up into some of its different systems, so that you can learn root words you might see together. There is no good way of remembering them except to memorize them and to relate them to terms you might already know. Got it? Ready, set, and go!

Skeletal System Root Words

These are root words that relate mainly to the bones and joints of the body. There are some that might seem to overlap or mean the same, but are used differently in medical language. Don't stress about these because you will eventually recognize when one term is more appropriate than another, seemingly related term.

- **Burso-:** This means "bursa," which is one of the many fluid-filled sacs near joints and tendons that help to lubricate these structures.
- **Carpo-:** This means "wrist." Your carpal bones are essentially your wrist bones.
- **Chondro-:** This means "gristle" or "cartilage." It is used in terms like *osteochondritis*, which is an inflammation of the bone and cartilage in the joints.

- **Dactyl-:** This is a root word meaning "fingers" or "toes." If you have *polydactyly*, it means you have too many fingers or toes on one hand or foot.
- **Ischio-:** This relates to the ischium or hip joint. Your *ischial tuberosity* is a bony protuberance near your hip joint.
- **Ossi- or osse-:** This is a general term to describe bone, or something that is "bony." The *ossification centers* in your bones are where new bone is made.
- **Oste-:** This also means "bone." If you are studying the field of *osteopathy*, you are partly studying diseases relating to the bones.
- **Ped-:** This is a term that means "foot" in anatomy. Your *pedal pulses* are the pulses you can feel on the top of your foot. A *pedometer* measures your step count when walking.
- **Pelv-:** This means "hip bone" or "pelvis." Your doctor might use a *pelvimeter* to measure the size of the pelvic bones when deciding if the birth canal is big enough.
- **Pleur-:** This is a root word meaning "ribs." Your *pleural cavity* is essentially the space bound by your ribcage.
- **Pod-:** This is another term meaning "foot." Who would your *podiatrist* be without this foot-related root word?
- **Cervico-:** This means your neck. The *cervical* area of your body is the area around your neck.

Muscular System Root Words

These are terms related to the muscles, limbs and joints. Many of the bone-related terms like *humerus* and *femur* are used commonly in anatomy, so they should be easy to remember over time.

- **Arthro-:** This means "related to the limbs or joints." The term *arthralgia* means to have pain in the joints in general.
- **Articulo-:** This also means something related to the joints. The term *articular surface* describes the surfaces that rub against one another in each joint.
- **Cost- or costo-:** This is a root word meaning "ribs." If you have *costochondritis*, it means that your rib joints are inflamed.
- **Humero-:** This is a root word meaning "shoulder." It explains why your *humerus*, or upper arm bone, is attached at the shoulder.
- **Lumb-:** This pertains to the back. When you say, "Oh, my lumbago!" it means that your back really hurts, and your *lumbar* area is the low back part of your spine.
- **Manu-:** This refers to your hand, and explains what you mean when you talk about *manual* labor being hand-related work.
- **Musculo-:** This simply means "muscle" and is the basis for the term *musculoskeletal system*, which refers to muscles and bones together.

Skin or Integumentary System Root Words

Your skin or integumentary system can involve things like your hair and nails as well. There aren't too many of these.

- **Trich-:** This is a root word meaning "hair." If you have *hypertrichosis*, for example, you are growing too much hair in the wrong places.

- **Capill-:** This is a term that also means "pertaining to hair." You have *capillaries*, which are tiny, hair-like blood vessels in your tissues.
- **Dermato-:** This is a basic term meaning "skin-related." Your *dermatologist* would just be called a "skin doctor" without this medical term.
- **Cutaneo-:** This is a term meaning "skin." Your *cutaneous nerves* are the nerves that directly affect the sensation in your skin.
- **Onycho-:** This is a term meaning "pertaining to the nails." If you have *onychomycosis*, it means you have a fungal nail infection.

Nervous System Root Words

Your nervous system includes your brain, spinal cord, nerves and sensations. There are several of these terms you will need to remember in order to study this system:

- **Acoust- or acou-:** This refers to hearing and is used to describe things like the *acoustic nerve*, which is responsible for your ability to hear well.
- **Aesthesi-:** This is a root word meaning "sensation." If you undergo *anaesthesia*, you are getting some type of local or general medication to numb either an area of the body or your whole body.
- **Algesi-:** This refers to pain or the perception of pain. The *analgesic* you take for a backache is something you are using to block the pain.
- **Auri-:** This means "ear." It's why, in anatomy, your outer ear is called the *auricle*.
- **Cerebello-:** This is a term used to describe the *cerebellum* in the back of your brain.
- **Cerebro-:** This is a medical term related to the brain itself. Your *cerebrum* is another word for your brain.
- **Encephalo-:** This is another term used to mean "brain." A common term using this root word is *encephalitis*, which simply means an inflammation or infection of the brain tissue.
- **Nerv-:** This means "pertaining to nerves" and is the basis for all terms relating to the *nervous system*.
- **Neuri-:** This is a term that is used for describing the nervous system. Your *neural pathways* are the paths your nerves follow in the brain and *peripheral nervous system*.

Endocrine System Root Words

There aren't too many of these terms. You have many endocrine glands that go by different names. You'll pick these terms up over time. Here are a couple you might need to memorize.

- **Thyr-:** This relates specifically to your thyroid gland, which is located in your neck. Your *thyroid hormones* are those made by your *thyroid gland*.
- **Adren-:** This is a root word meaning "adrenal glands," which breaks down further because "ad" means "on top of" and "ren" means "related to the kidneys." You can remember where the adrenal glands are because they're on top of the kidneys.

Circulatory System Root Words

The circulatory system terms are all related to the blood, the heart, and the blood vessels themselves. There are a few root words and some suffixes and prefixes that all relate in some way to the circulatory system. Here are some to remember:

- **-emia:** This means "related to the blood or blood condition." For example, *polycythemia* means too many blood cells in the blood.
- **Angio-:** This means "blood vessel." If you've ever had to have an *angiogram*, you'll know this is an imaging test of the blood vessels.
- **Arterio-:** This means "pertaining to an artery." An *arteriogram* is a subtype of an *angiogram* that specifically looks at the way the arteries look on the inside.
- **Atrio-:** This means "pertaining to the atrium," a small chamber in the heart. Your *atrioventricular valves* are between the *atrium* and the ventricle.
- **Corono-:** This means "related to the heart" and involves terms like the *coronary arteries,* which are the blood vessels that cover the heart.
- **Hemat-:** This is a root word meaning "blood." There are a lot of terms using this root word, including *hematology,* which is the study of blood and blood diseases.
- **Hemangio-:** This is a term that means "blood vessels." You might see a tumor called a *hemangioma,* which is a blood vessel tumor on the skin made from a tangled mess of blood vessels.
- **Phlebo-:** This is a root word that specifically means "veins." Your *phlebotomist* is the person who draws blood from your veins. If you have *phlebitis,* it means your veins are inflamed.
- **Sangui-:** This is another term meaning "blood." If you happen to *exsanguinate* (and I hope you don't), it means you lose all your blood, usually from a major traumatic event.
- **Thrombo-:** This means "pertaining to a blood clot." If your *phlebitis* is instead called *thrombophlebitis,* it means you have a blood clot inside an inflamed vein.
- **Vaso-:** This is yet another term for "blood vessel." *Vasovagal syncope* causes you to faint. One reason for vasovagal syncope is that your blood vessels are too dilated and your blood pressure gets too low.
- **Ven-:** This is a term that means "vein." If you have a *venogram,* it is a type of angiogram that looks specifically at the inside of the veins.

Respiratory System Root Words

The respiratory system involves several terms related to your lungs or airways. These are the ones you should remember:

- **Broncho-:** This is a term meaning "bronchial tube." It is used in the term *bronchoscope,* which is an instrument used to look inside the bronchial tubes.
- **Bronchio-:** This term also means "bronchial tube" and is the root word for several terms linked to your *bronchi,* or airways.
- **Bronchiolo-:** This is a root word meaning "small airways." Your *bronchioles* are the smaller airways in your lungs.
- **Aero-:** This is a Greek term meaning "gas" or "air." When you do *aerobic exercise,* it means you are exercising with air or with a lot of breathing.

- **Pneumo-:** This is a basic root word meaning "lungs." When you use the term *pneumonia*, you are referring to a lung infection.
- **Pulmono-:** This is a term that also means "lungs." This is where the term *pulmonologist*, or lung specialist, comes from.

Urinary System Root Words

Your urinary system usually means things related to your kidneys or urine. Some of these terms blur into the reproductive system terms we will talk about next. Here are some common urinary system root words:

- **Nephro-:** This is a root word meaning "kidneys." If you have *nephritis*, it means that your kidneys are inflamed or infected.
- **Ren-:** This is a root word that also means "kidneys" but is Latin in origin. Your *renal arteries* are the blood vessels that supply blood to the kidneys.
- **Ur-:** This is a short root word that means "urine" or "pertaining to the urine." If you are talking about *polyuria*, it means you are urinating too much or too often.
- **Vesico-:** This is a root word that means "bladder." Your *vesicourethral valve* is the valve between the bladder outlet and the urethra, through which urine passes outside the body.

Reproductive System Root Words

There are reproductive system root words related to embryology and sexual organs, both female and male. These will be familiar to you some of the time, but not so obvious in other examples given. Let's look at these:

- **Episio-:** This means something related to the pubic region or loins. In female reproductive medicine, an *episiotomy* is when the doctor makes an incision in the pelvic floor to widen the birth canal during childbirth.
- **Balan-:** This is a term that refers to the glans penis in men. Men who have *balanitis* have an infection or inflammation of the tip of their penis.
- **Colpo-:** This means something related to the vagina. A doctor might use a *colposcope* to take a close look at the inside of the vagina.
- **Gono-:** This means "semen" or "seed." The *gonads* are the male and female reproductive organs, and *gonorrhea* is a sexually transmitted disease in both men and women.
- **Gyno-:** This is a word that simply means "woman" and explains why a *gynecologist* really only deals with women's sexual health.
- **Mammo-:** This is a basic root word that means "breast." A *mammogram* is an imaging test or X-ray of the breasts.
- **Masto-:** This also explains things related to the breast. A woman with *mastitis* has a breast infection.
- **Metry:** This is a root word that means "pertaining to the uterus." The *myometrium* of the uterus is a muscle wall and one of the layers of the uterus itself.
- **Oo- or ovo-:** Both of these are root words that mean "egg" or "eggs." An *oocyte* is a female egg cell, and the *ovaries* are where the eggs in women are made.

- **Ovario-:** This means "related to the ovaries." Your *ovarian ligament* is one that's connected to the ovaries in women.
- **Salping-:** This is a root word that means "fallopian tubes." A woman having a *salpingectomy* is having her fallopian tube removed.
- **Spermato-:** This is a root word referring to sperm cells or semen. *Spermatozoa* is another word for sperm cells.
- **Vagin-:** This is a term that means "relating to the vagina." A woman with *vaginitis* has an infection or inflammation of her vagina.

Gastrointestinal System Root Words

Everything from the mouth to the anus and all the internal organs of the abdomen fall into this category of medical terms related to the GI system. Some of these have overlapping meanings that you should try not to become too confused about. Here are the terms to remember:

- **An-:** This can mean "anus" and is the term you use when you talk about having an *anoscopy exam*, which is a really close look at your anus with a special instrument.
- **Hepat-:** This is a root word that means "liver." If you have *hepatitis*, it means there is an infection or inflammation of your liver.
- **Laparo-:** This is a term that means "abdominal wall." A *laparoscopy* is a term used to describe a test in which an incision is made in the abdominal wall in order to look inside the abdomen.
- **Pepsia:** This is a suffix that means "digestive tract." If you have *dyspepsia*, you have indigestion from pain in your digestive tract.
- **Phagy-:** This just means "eating" or "feeding on." The term *phagia* is similar to this. If you are suffering from *dysphagia*, you experience painful or difficult eating.
- **Pharyngo-:** This means "pertaining to the throat." When you have *acute pharyngitis*, you probably have a basic sore throat.
- **Procto-:** This means "rectum" (or perhaps "anus"). A *proctoscope exam* is slightly different from an anoscope exam because the proctoscope looks further up into the rectum than the anoscope.
- **Rect-:** This basically means "rectum" and refers to the part of your intestinal tract just before your anus. You can guess where a *rectal suppository* might go if your doctor recommends you have one.
- **Abdomin-:** This relates to the abdomen itself. If you have an *abdominoplasty* procedure, you are having a "tummy tuck."
- **Cholecysto-:** This means "gallbladder." So if you have *cholecystitis*, you are experiencing an infection or inflammation of your gallbladder.
- **Colo-:** This means "pertaining to your colon" and is where the term *colorectal cancer* comes from.
- **Duodeno-:** This refers to the duodenum, or the first part of the small intestine.
- **Entero-:** This is the root word for "intestine." When you have *gastroenteritis*, you have an infection of your stomach and intestines.
- **Esophago-:** This is a root word that means "gullet" and relates to the tube between your throat and your stomach.
- **Gastro-:** This is the term used to describe your stomach. The word *gastritis* means there is inflammation or an infection of your stomach.

- **Geusia:** This is a fancy term for "taste." A person who has *dysgeusia* simply has a bad taste in their mouth.
- **Ilio-:** This is a root word meaning "ileum." Your *ileum* is the last part of the small intestine.
- **Pancreat-:** This is a root word that means "pancreas."

Let's Wrap this Up!

If this chapter seemed challenging to you, you are probably not alone. There are some terms we've talked about that you may have never heard of before, and a few you might never hear again. There are hundreds of medical terms and we didn't cover them all; so if this sort of thing really excites you, you can get a book on medical terminology and knock yourself out with a lot more terms.

And now for the grand finale! Let's go back to the **EXTRA CREDIT** pop quiz we talked about in the beginning of this chapter. What happens in an endoscopic retrograde cholangiopancreatography test, or ERCP? If you piece it all together, you might guess that it is an imaging test that uses an internal scope to look at the structures of the gallbladder, pancreas and *bile ducts*, by injecting dye into the opening of these structures in the duodenum. Yeah, that's really what it is, and no wonder they just call it an ERCP. If you guessed this one even close to correctly, you're well on your way to being fluent in medical terminology. Congratulations!

To Sum Things Up

Most of the medical terms we use today would have made more sense to those who once spoke ancient Greek and Latin. That's because those languages are the origins of the medical terms we use today. Partly this is because some of these anatomical terms have existed as far back as ancient times, but most are more modern. By convention, they're still named with Latin and Greek terminology in mind.

Most medical terms have a prefix, root word and suffix. These are word parts that each play a role in what the word means. After you have studied these word parts, you should be able to figure out (at least roughly) what an unfamiliar medical or anatomical term means.

Chapter Three: Your Cells and Tissues

Your own human body has more than thirty trillion cells in it! Cells of the same type make up your tissues, and tissues that participate in the same function in the body form your organs. Organs can be made of more than one tissue type.

Different organs each have a job in the body, and work together to make what we call the physiology of the body. The anatomy of your cells, tissues and organs are like the car parts under your car's hood. The physiology is what happens when you turn your car on, use up gas, and drive your car somewhere. Without your car parts and energy systems working together, your car would be useless. This is exactly the same thing happening in your body.

Basic Cells and What's Inside Them

A long time ago, right after your mom's egg cell and your dad's sperm cell united, you were just a single-celled structure called a zygote. Within seconds, you began to divide rapidly, with one cell becoming two cells, then two cells becoming four cells, and so on. It didn't take long before you were made of hundreds, thousands, then millions of cells. These cells began to differentiate and specialize, so that you were no longer just a blob of cells but instead a small being that contained cells of many types.

How can we study the basic cell structure in the human body, knowing that your cells are all different from one another? It's because there are a lot of human cell features that are the same, no matter what cell you're studying. Later, you'll learn why each cell of the body is very different from the cell next to it, even though each cell has the same genetic material inside it. It wouldn't make sense for every cell in your body to do the same thing.

Basic Cell Structure

When we talk of the basic cell structure of the human body, it's a lot like describing your carburetor and your fuel line as if they were the same thing. You'll just have to trust that, while each cell is different, it's the similarities that are more important when studying human and animal cells. In this section, you'll learn about what makes the cells of your body similar to one another and not so much about what makes them different.

Cells are often called the basic units of life. These are the individual, membrane-bound structures inside your body, and they have a lot going on inside them. Even though humans contain trillions of cells, there are gazillions more organisms on Earth that live as single-celled animals.

Fun Factoid: Antonie van Leeuwenhoek was a Dutchman in the 1600s who made his own microscopes by grinding glass lenses. He was the first to see small cells, including bacterial cells and animalcules, which are what we now call protozoa. *He is called the father of microbiology because of his early discovery of many types of cells.*

Like humans, the cell in single-celled creatures is kind of a self-contained unit. The difference between single-celled organisms and your cells is that your cells work together to help you function. Unlike a single-celled organism, you really are more like a car engine that needs all the parts working together to help you stay healthy.

This figure of a cell shows that, even though you might not see anything clearly under a microscope, there are more than a dozen different internal cell structures called *organelles*. This basically means "tiny organ." Technically, to be called an organelle, the structure must have its own membrane around it, but most people who study the anatomy of the cell call every structure an organelle, even if it doesn't have its own membrane.

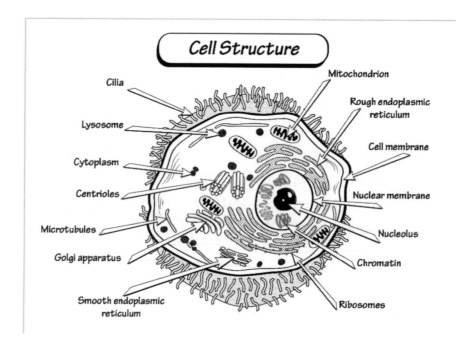

What does each organelle do in order to help the cell function? Let's look at the different organelles more closely to see what they look like, and what their function is inside the cell. If you don't think you get it after reading this chapter, you'll catch up later as we cover some of these organelles in greater detail.

The Cell Membrane

The membrane is the outer boundary of the cell. Without it, all your cell contents would spill out. The interesting thing about this membrane is that it is *semi-permeable*. What this means is that it lets some things enter and exit the cell freely, while other things don't get in at all, or can only enter the cell through specialized means.

In humans, the cell membrane is made from *phospholipids*, cholesterol and proteins. The phospholipids make up the majority of this membrane. These are unique molecules that have one end that likes water, called the *hydrophilic* end, and one end that doesn't like water, called the *hydrophobic* end.

Since the environment inside and outside the cell is all water-based, the phospholipid membrane lines up in a way that keeps the hydrophobic ends away from water and the hydrophilic ends next to water. In order to do this, the phospholipids form what we call a *phospholipid bilayer*. The figure shows what a phospholipid and a phospholipid layer look like:

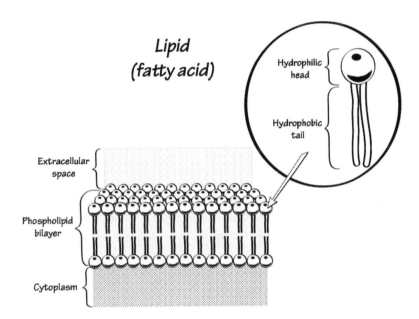

This phospholipid layer is fluid like the ocean. Floating in this bilayer are cholesterol and proteins. They don't stay in the same place. Unlike debris, however, the cholesterol and proteins have a real job to do. Proteins help to make channels for substances to pass through the membrane, and cell receptors that tag the cell on the outside. These tags can identify the cell or can provide attachment points for other molecules to stick to the cell membrane. Once this happens, there can be a change in the biochemistry of the cell without having a molecule actually get inside the cell. This figure shows what this phospholipid raft in the cell membrane looks like:

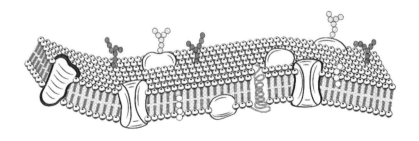

We will talk more about how things get inside and outside of the cell in the next few chapters. For now, you should understand the idea that this floating raft has important structures in and on it that help the cell function in many ways. Cholesterol also floats on this raft inside the membrane around the cells. Only animal cells have cholesterol in their cell membrane, which keeps cells from freezing in low temperatures.

Fun Factoid: As you will learn later in the book, cholesterol is not the bad guy in your body. Yes, cholesterol plaques can cause you to have a heart attack if the numbers are too high, but cholesterol also makes up the myelin sheath that surrounds your nerves. Without your myelin sheath, your nervous system simply can't function very well. Just ask anyone with multiple sclerosis, which is an autoimmune disease where your immune system attacks and damages the nerves' myelin sheaths.

The cell membrane is actually quite ingenious. There are different ways that substances can cross it in the cell. One of these is called *diffusion*. This is just the passage through the membrane of small hydrophobic cells, from areas of high concentration to areas of low concentration. Gases like oxygen and carbon dioxide, and small substances like alcohol, can simply cross through the membrane without help.

Facilitated diffusion requires the use of carrier proteins or channel proteins to allow polar molecules that can't otherwise easily cross the membrane to get through. Channel proteins have holes in them to allow molecules to drift through the membrane, while carrier proteins will change shape when in contact with certain molecules to allow the molecules to pass. Small proteins, glucose and ions will pass through the membrane this way.

There are a couple of really cool mechanisms used in the transport of small molecules and ions against a concentration gradient. What this means is that these molecules must be pulled uphill from an area of low concentration to an area of high concentration. These are called *symport* and *antiport* mechanisms. It works by having one ion, usually sodium, pass through a channel normally from an area of high concentration to an area of low concentration. This helps draw another ion through the same channel in the opposite direction in an antiport system, or in the same direction in a symport system. This is shown in the figure below:

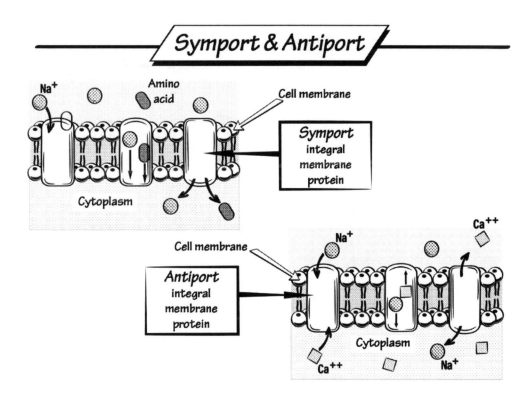

Water gets through the cell membrane in two ways. It can drift through the membrane through the process of *osmosis*, which is like diffusion but just for water. Water flows from a dilute solution to a concentrated solution across a membrane in order to even out the amount of water on both sides. There are also water channels called *aquaporins* that help water get through the membrane faster.

Finally, there are times when you just need a boost of energy to get a molecule across the membrane in an uphill situation, from an area of low concentration to an area of high concentration. This is called *active transport*. It uses *ATP energy* from the cell to drive a pump that drags molecules across the membrane against its concentration gradient. There is one common pump in the membrane that does this, called the *sodium-potassium ATPase pump*. It uses energy to pump sodium out of the cell and potassium into the cell. This is really important because it also makes the electrical charge in the cell become more negative compared to the outside. This is what the sodium-potassium pump looks like:

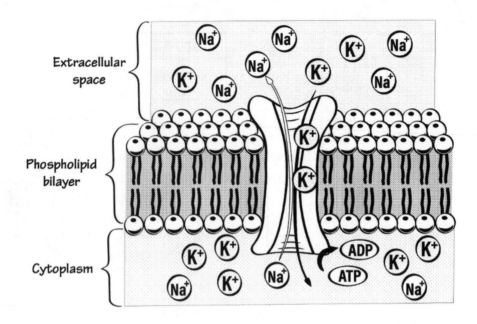

The Nucleus

The *nucleus* is a major organelle in the body. It has a double membrane around it with two separate lipid bilayers, and nuclear pores puncturing the membrane so that structures inside the nucleus can get out. Inside the nucleus is the genetic instructions for the cell. It's like the central processing unit of a computer that carries all the necessary information the cell needs to do its job. This genetic information in humans (and all animals) is called *deoxyribonucleic acid*, or DNA. DNA in your cell is found in 46 separate segments called *chromosomes*. The codes for instruction on how to make cell proteins are determined by how the segments of DNA are arranged.

DNA on a chromosome is organized into *genes*, or segments, that each code for a protein in the cell. You would think that every piece of DNA is useful to you, but only one percent of your DNA actually codes for proteins. About 25 percent of your DNA codes for regulatory factors, which are just used to make sure the right genes get turned on in the cell. The rest of your DNA seems to be junk as far as researchers can figure out. This figure shows what the nucleus looks like:

Cell Nucleus

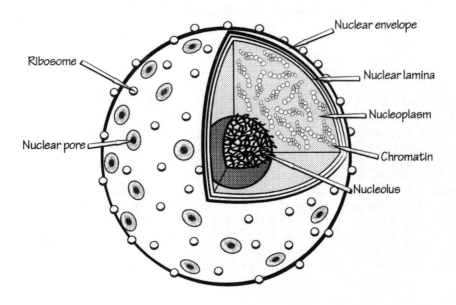

Inside the nucleus is a structure called the *nucleolus*, which doesn't have its own membrane. This is just an area of the nucleus that is segregated for making *ribosomal RNA*, which is one of several types of RNA—another nucleic acid similar to DNA—that strictly makes the protein factories in the cell called *ribosomes*. Without a normal nucleolus, proteins couldn't get made, because this factory is shut down in the cell.

The watery stuff inside most of the cell is called the *cytosol* or *cytoplasm*. It isn't completely watery but is more like a gel that suspends all the organelles inside the cell. It has some sort of structure and compartmentalization. There are *filaments* of different types, that act like tiny roadways to allow the movement of substances in the cell. These filaments also form the shape of a cell, much like the tent poles of a tent hold up the tent itself.

Endoplasmic Reticulum: A Cluster of Membranes

The *endoplasmic reticulum* is a membrane structure near the nucleus in the cell. It looks like a stacked series of membranes that are actually attached to the nuclear membrane. There are two types of endoplasmic reticulum. Rough endoplasmic reticulum, or RER, is called rough because it looks that way under the microscope. It is studded with ribosomes where proteins are made. Once the proteins are made, the endoplasmic reticulum is where they are processed.

There is also smooth endoplasmic reticulum, or SER, which looks smooth because it has no ribosomes on it. It makes lipids instead of proteins and then processes them for other parts of the cell. This figure shows what the endoplasmic reticulum looks like:

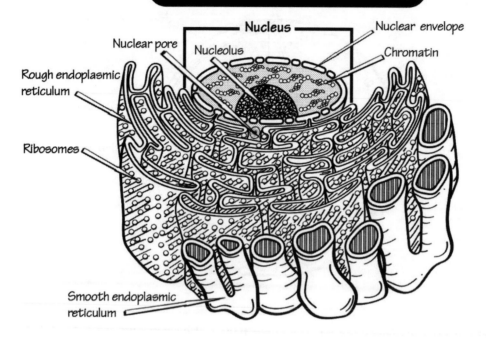

Nucleus and Endoplasmic reticulum

Nucleus

Nuclear pore

Nucleolus

Nuclear envelope

Chromatin

Rough endoplasmic reticulum

Ribosomes

Smooth endoplasmic reticulum

The Golgi Apparatus: The Cell's Post Office

Once proteins get made and processed, they need somewhere to go. Have you ever wondered how these proteins know where they are most needed? There is a *Golgi apparatus*, or *Golgi body,* near the endoplasmic reticulum, which is basically the post office of the cell. Molecules of all types enter a series of flattened sacs at one end of the apparatus. Once inside, they get labeled and collected into *vesicles*, which are small membranous sacs. They leave the other end of the Golgi body fully stamped and packaged, like something you'd mail at the post office.

On each package of proteins and other substances are the instructions that tell the cell to send the package somewhere inside the cell or perhaps outside the cell. The cytoskeleton filaments in the cytoplasm are how these vesicles go from place to place. This figure shows you what the Golgi body looks like:

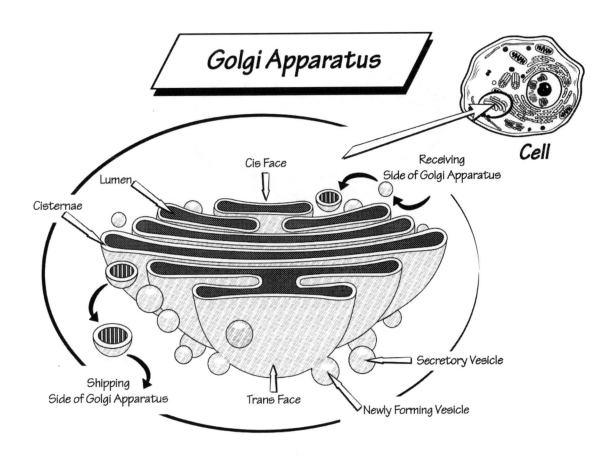

Mitochondria: Your Cell's Power Plants

Mitochondria are the energy structures inside each cell. They are like the piston in your car that keeps on making energy for the body as long as there's enough gas in the tank. The food you eat makes up that fuel for your cells. Mitochondria turns this fuel into useable energy by making ATP energy molecules, used to drive almost all major biological processes in the cell. This figure shows you what the mitochondria look like:

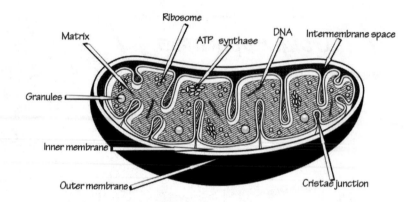

Ribosomes Make Proteins

Ribosomes are the protein-making factories found on the rough endoplasmic reticulum. These are mostly made from a special type of RNA molecule that forms a complex that directs the DNA messenger molecule coming from the nucleus. This messenger molecule is called *messenger RNA*. It literally carries the message necessary to make protein in the ribosomes. Which protein gets made depends on what the actual messenger RNA message is telling the ribosome to do.

Lysosomes and Peroxisomes

Lysosomes are membrane-bound organelles in the cytoplasm. They aren't very big, but they are powerful. They are like incinerators that take waste molecules in the cell cytoplasm and, using strong breakdown *enzymes*, destroy larger molecules in order to make small molecules. These small molecules often get recycled in order to make new products for the cell. *Peroxisomes* are like lysosomes, but they use powerful *oxidizers* to break down molecules instead of the enzymes used by lysosomes.

Fun Factoid: You can remember what peroxisomes do by taking apart the word to make peroxi/some. The first part of the word is the same root used in the word peroxide, or H2O2. This is an antiseptic that you might use on a cut or sore, but in reality it's also a strong oxidizer found inside peroxisomes. The last part of the word means body. *By remembering that peroxide is inside these bodies, you'll remember that they destroy things by oxidizing them.*

Tissues in the Body

One cell in your body is a great idea, but many cells of the same type working together can really get the job done in ways one cell by itself can't do. Let's look at the different tissue types inside the human body. While you'd think there were a lot of different types of human body tissue, there are really only four types. Within these types, however, are cells that are unique to certain bodily areas. These are the different major types:

Epithelial cells are mainly your boundary cells. They line all the major body cavities, like the respiratory tract, bladder and GI tract. Your skin is made from epithelial cells. You need these to keep stuff from getting into your body that might be dangerous. These often have two sides without any major blood or nerve supply. The *apical side* faces the interior of the GI tract, bladder or respiratory tract, and the *basal side* faces away from that. They are lined up on a protein sheet called a *basement membrane* and are connected by *tight junctions* between them that keep substances from passing through their layers. This figure shows what epithelial cells look like:

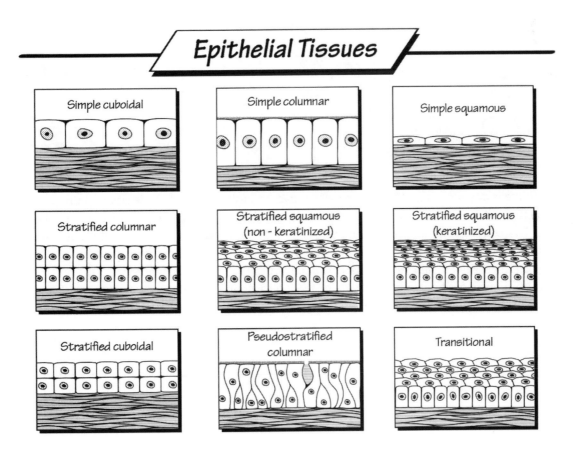

Connective tissue is basically structural tissue, like the steel skeleton that holds up your house. These are cells that have protein fibers near them that support most other tissue types in the organs of the body. Common fibers you'll see in connective tissue are *elastin* (which makes tissue elastic), *collagen*, and *reticular tissue* (which is made from loosely-arranged connective tissue fibers). Blood is considered a connective tissue, even though it looks nothing like other types of connective tissue. Fatty tissue (or *adipose tissue*), cartilage and bone are also considered connective tissue. This figure shows what adipose tissue looks like:

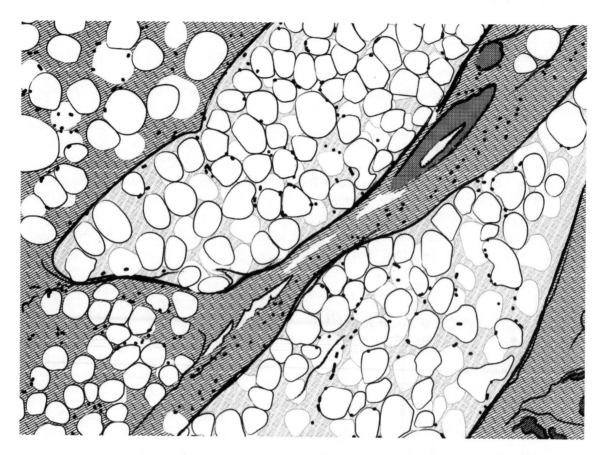

Muscle tissue is any moveable tissue that helps move body structures. Muscle tissue is unique because it is contractile and moves itself. When you think of muscle tissue, you probably think of your skeletal muscles. There are cardiac and smooth muscle cells in addition to skeletal muscle. Cardiac muscle cells are found in your heart and help it to pump. Smooth muscle is involuntary (as is cardiac muscle because it moves on its own). It's the kind of muscle found in your blood vessels, bladder, uterus and digestive tract. Smooth muscle is how your food gets propelled through your GI tract without you having to think about it. This figure shows what the different muscle cells look like:

Types of Muscle Cells

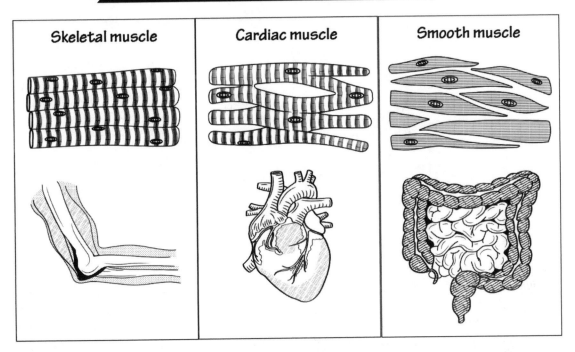

| Skeletal muscle | Cardiac muscle | Smooth muscle |

Finally, there is nervous tissue. These are actually made from two types of cells. There are *neurons*, or nerve cells, that have cell bodies with multiple extensions called *axons* and *dendrites*, that pick up chemical signals from other nerve cells. These chemical signals get transferred into an electrochemical signal, that sends messages to muscle cells and other nerve cells through a unique but extensive electrochemical system in the body. There are also supportive cells called *glial cells* that do things like make the myelin sheath around nerves, and protect the nerves from damage by infectious or inflammatory cells.

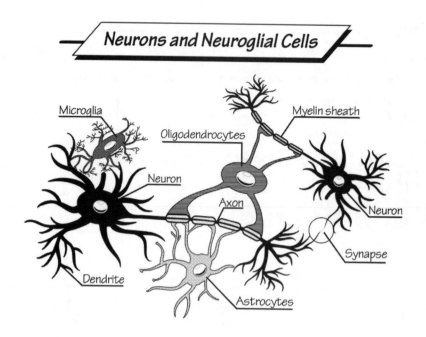

Neurons and Neuroglial Cells

It takes tissues of different types working together to make a single organ. In the liver, for example, there are *parenchymal cells*, or liver cells, that are supported by connective tissue cells. There are epithelial cells that line the ducts, and nerve tissue that send signals to the liver to control its function. In the blood supply to the liver, there are blood vessels with smooth muscle cells in its walls that control the blood flow to the liver. It takes all of these different cell types to make the liver a whole organ.

As you can see, like a car engine, the cells need to function independently, and the different cells need to form tissues that all together make the whole organ function the way it should. Taking it one step further, all of the organs are the different parts of the whole car, and act in tandem to keep the car going on the road. On a much smaller level, this is how your body works as well.

To Sum Things Up

Your body starts as a single cell that divides rapidly. As an embryo grows, the cells differentiate into the four different major cell types. These cell types together form the different tissues and organs inside your body. In this chapter, we talked about what the different cell types are and how they are unique from one another.

While the cells of your body have different jobs, there are similarities that help unite all of your cells as being some type of animal cell. We talked about the importance of the cell membrane, and about each of the organelles that function together to get the jobs done inside the cell, so that it can be a living factory inside your body. While you have trillions of cells, there are many more single-celled animals on Earth who have cells that are very similar in many ways to our own.

Chapter Four: Crossing the Cell Membrane Using Cellular Transport

Without the actions of the cell membrane, a cell would be like a balloon with no chance of anything getting in or out. Cells are instead dynamic structures, continually taking on nutrients and giving off waste products. For example, all your cells need oxygen to be able to metabolize nutrients. When oxygen is part of this metabolic process, carbon dioxide is made as a byproduct. You need the nutrients to fuel your cells, and you need a way for oxygen to get in and for carbon dioxide to get out.

Another important thing that happens because of cell transport is that ions like sodium, potassium and chloride enter and leave the cell all the time. There is a huge difference between the concentrations of these ions when you compare the inside of the cell to the outside. This means that there must be ways to maintain these concentration differences. This table shows you some of these differences:

Ion Type	Concentration in the Cell	Concentration Outside the Cell
Sodium	10 millimoles per liter	140 millimoles per liter
Potassium	140 millimoles per liter	4 millimoles per liter
Chloride	4 millimoles per liter	100 millimoles per liter
Calcium	0.1 millimoles per liter	2.5 millimoles per liter
Magnesium	30 millimoles per liter	1.5 millimoles per liter
Bicarbonate	10 millimoles per liter	27 millimoles per liter
Glucose	Less than 1 millimoles per liter	5.5 millimoles per liter

If you put a salt or sodium chloride tablet on one side of a totally permeable membrane and a potassium chloride tablet on the other, the tablets would dissolve and over time the concentrations of potassium, sodium and chloride would be equal on both sides. The cell membrane is semi-permeable. Some things get in and out easily, while others just don't cross the membrane at all without a lot of help. This figure shows what a totally permeable membrane would do:

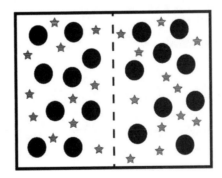

Since all molecules of gas, liquids, and even solids are always in motion as long as the substance is above 0 degrees Kelvin (also called absolute zero), energy must be used to move molecules from a higher concentration to a lower concentration and to keep them that way. Eventually, if no energy is put into your body's cells, the different molecules would reach a state of equilibrium, where they would be sent to various parts of the body and the concentrations inside and outside the cell would be the same.

As you can see from the table, this doesn't just happen in your body. The reason? Energy. We put energy into membrane systems that allow ions and larger molecules to get pumped across the cell membranes regularly. This is what allows the inside cellular environment to be so different from the outside.

You will see in this chapter the ingenious ways cell membranes have developed over the course of time to keep things where they belong. Some of these mechanisms require no energy input at all, while others need a continual input of energy. For example, you need to know that keeping ions where they belong involves creating an electrochemical gradient across the cell.

A *gradient* is basically a difference. When you go from the bottom of a hill to the top, you are going up a height gradient. When storms happen, it is because of *pressure gradients*, or pressure differences in the atmosphere. Some gradients are subtle, while others are obvious. When you see a sign that says "steep grade ahead," it basically means the gradient will be steep.

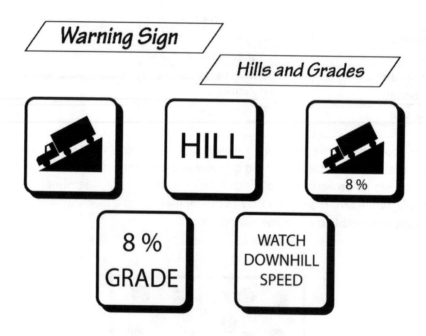

Electrochemical gradients are part electrical and part chemical. Remember how ions are charged particles? The cell membrane arranges its transport mechanisms to keep the resting cell more negatively charged inside the cell membrane compared to the outside. This becomes really important in nerve and muscle cells that use the electrochemical gradient to allow messages to be passed along a nerve cell, and to allow a muscle cell to contract.

One ingenious way this is done is through a pump called the sodium-potassium ATPase pump. It exchanges sodium for potassium across the membrane. But this is the key feature to remember: It is not a 1:1 exchange. For every three sodium ions passed outside the cell, two potassium ions are passed into the cell. Both ions are positively charged with one unit of charge each. This exchange means more negativity inside the cell and more positivity outside the cell. This figure shows you what it looks like:

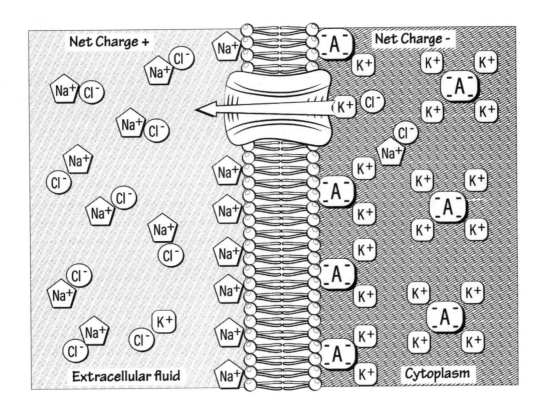

Osmosis

Osmosis is all about the movement of water. It certainly could apply to any solvent in nonliving systems, but in humans it's all about water flow. Osmosis makes the best sense if you think of water as, well, not really liquid, but instead just a bunch of water molecules bumping around randomly into one another, and into anything else they come in contact with. This bumping around is normal with any liquid, because liquid molecules have energy and they like to jiggle all over in a space or container. This jitteriness is called *kinetic energy* but you can think of it like water hyperactivity. This figure shows water in liquid form:

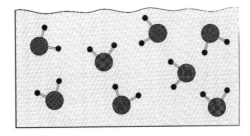

In living systems, water bumps into itself, into other molecules, and into the cell membranes because of kinetic energy. Concentrated solutions with a lot of ions and molecules in them have more kinetic energy than dilute solutions. As in any energy system, the tendency of energy is to flow from areas of high energy to areas of low energy in order to even things out. It's similar to dumping hot water into one side of your tub and cold water into the other side. Hot water is high energy and flows into the colder part of the tub. In short order, you have a tub of evenly distributed warm water.

Unlike your bathtub, living systems have membranes to prevent the free flow of molecules anywhere they would otherwise want to go. To understand best how water gets from one side of a membrane to another, think about what happens if you forget to water your tomato plants for a while. In a few days, they look all wilted and near death. When you try to revive them by watering them, a miracle occurs! Within a couple of hours, your tomatoes look healthy again.

Plant wilting happens because the cells of the plant get dehydrated. Water leaves the cells and the cells lose their *turgor pressure*. Turgor pressure is the inside pressure of the cell. It's also referred to as *hydrostatic pressure*. It's what makes fruit look luscious and what keeps a cell nice and plump. There are tough cell walls around each cell plant, but this isn't enough to keep the plant from wilting.

When a plant cell loses water, it is more concentrated with respect to the non-water molecules inside the cell. There are more ions and small molecules compared to how much water is in the cell. When you water it, you put in pure water, which isn't concentrated at all. There's a lot of water but few ions in your watering can. The water you give the plant then moves from the less concentrated areas outside the cell to more concentrated areas inside the cell. This will plump up the cells and make your tomato plants look healthy again.

These randomly moving jittery water molecules bounce more against one another when the concentration of water in the area is greatest. This might seem confusing until you think of water as having a concentration, just like the solutes in it. High water concentrations translate to dilute solutions with respect to the solute. Remember that the solutes are all of the non-water substances dissolved in the solution.

Dehydrated cells have a concentrated solution of ions or other molecules, and are low in their water concentration. Water will go then from its high concentration (dilute state) to its low concentration (dehydrated state). This figure shows osmosis across a semipermeable membrane, permeable to water but not to the ions in the water:

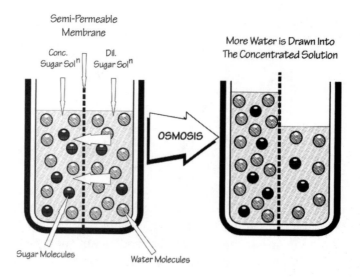

This leads us to a couple of other important concepts. The first is called *osmolarity*, which is the concentration of all the dissolved solutes in a compartment per liter of solution. When we think of dissolved solutes, we have to think of all solutes—like sodium, chloride, potassium, sugars and proteins. Salts are doubly counted, because those like sodium chloride are a combination of one sodium ion and one chloride ion. High osmolarity is roughly the same as high concentration, even though they're technically different.

The other concept you should know about is called *tonicity*. Tonicity is a relative term that describes the water concentration between two compartments or the measurement of the effective osmotic pressure gradient. This leads to terms like *hypotonic*, *hypertonic* and *isotonic*. If you think of a cell as being a compartment with its own osmolarity and tonicity, compartments outside the cell can have the same, greater, or less solute concentration or tonicity.

When you think about it, tonicity is a term of comparison between two separate compartments, while osmolarity is an absolute number that doesn't really compare itself to the osmolarity of the compartments separated from it. This is the main difference between these two concepts.

Hypotonic solutions outside the cell are more dilute and have a lower osmolarity. Isotonic solutions outside the cell have the same osmolarity as inside the cell. Hypertonic solutions are more concentrated and have a higher osmolarity compared to inside the cell. What happens to the cell because of osmosis if it's placed in these environments? This figure describes exactly that:

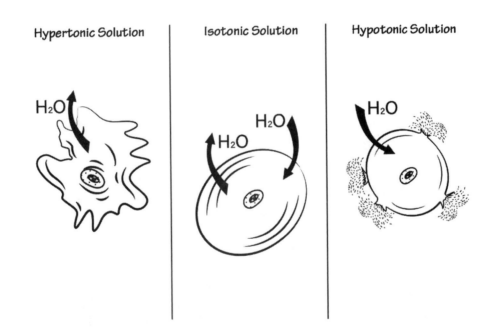

The cell in a hypertonic solution will cause water to rush out of the cell membrane, shrinking it. The cell in a hypotonic solution will cause water to rush into the cell, bursting it. The cell in an isotonic solution will have water entering and leaving the cell equally, so no change in cell size will be seen.

Why and when is this important in human physiology? In the kidneys, water is filtered out of the blood along with ions and small molecules. If none of it were reabsorbed, not only would you pee a lot, but you would need to drink a lot of water.

In the tubules of the kidneys, water goes from the inside of the tubules back into the bloodstream because of osmosis. Tubules of the kidneys are impermeable to ions but not to water. Through a unique mechanism, the osmolarity of urine is dilute at 300 milliosmoles per liter in the beginning but is more concentrated at 1200 milliosmoles per liter after osmosis has occurred. This figure shows what that looks like:

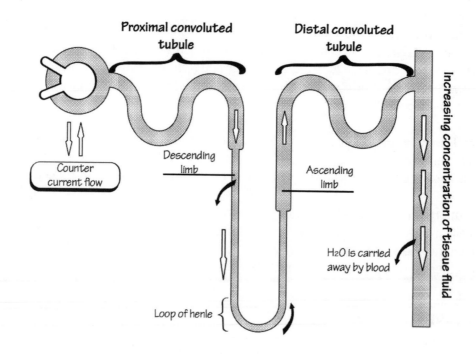

To sum things up: Osmosis is the net spontaneous movement of water through a semipermeable membrane, from a low concentration of solute to a higher concentration of solute. It does this in your body in order to achieve homeostasis, or sameness, inside your cells.

Osmotic pressure is the pressure of an enclosed biological space, that depends on the concentration of the solute within that particular space. It's also something you can measure. Osmosis is related to the osmotic pressure in a single space. Osmotic pressure drives water across the membrane into another space with a higher osmotic pressure. The same osmotic pressure inside and outside the cell means that there is no net movement of water across the cell membrane.

Opposite to the osmotic pressure between two system is the hydrostatic pressure. This is more like the kind of pressure you see in your daily life. For example, a hose with a high hydrostatic pressure will squirt water out of it rapidly, because the water in the tube fills the hose under high pressure. A low hydrostatic pressure would be a situation where water trickles out of the hose. There isn't a lot of pressure bulging the hose, so the water drips out of it more slowly. In blood vessels, if the hydrostatic pressure is large in the vessel, the tendency is for water to squeeze through the blood vessel walls. It will be opposed by low osmotic pressure outside the blood vessel, which feels like it has enough water in it already, so it really doesn't want any more.

Diffusion

Diffusion is a natural physical process that often has nothing to do with your body. For example, what happens if a woman who clearly has put on too much perfume walks into a room? Within a few minutes, the whole room smells just like her. This happens because the perfume molecules on her become airborne and, through diffusion, travel around the room so they are more or less equally distributed.

Diffusion is nothing more than the travel of molecules in water or air from an area of high concentration to an area of low concentration. Like the jittery water molecules we just talked about, other molecules have kinetic energy and bump into one another. Molecules of a higher density bump into more of each other than molecules of a lesser density. This naturally causes a spreading out of the molecules so they even each other out.

In living things with cell membranes, there is a barrier to diffusion with some molecules but not others. This barrier is the cell membrane itself. Remember that the cell membrane is hydrophobic, so molecules that are polar or hydrophilic won't be able to diffuse across it. Nonpolar molecules that are hydrophobic themselves won't see the cell membrane as a barrier at all, so diffusion across it is possible.

Which molecules will cross over and which will not? All hydrophobic molecules will do this easily. Hydrophobic molecules aren't common in human physiology, but lipids like steroid hormones and other fatty substances can diffuse across the membrane. Gases like oxygen and carbon dioxide are small and nonpolar, so they have no problem diffusing across the membrane. Water does not technically diffuse but passes through the membrane through osmosis, which is a kind of diffusion.

This next figure shows what happens all the time in almost all the cells of your body, as capillaries rush by to deliver oxygen to them. Oxygen diffuses from the outside of the cell to the inside, while carbon dioxide goes the other way. Because your tissues are using up oxygen all the time, the concentration of this gas is low inside the cell, so oxygen can diffuse from the outside (where the concentration is high) to the inside. The reverse is true for carbon dioxide, which gets made inside the cell, so its concentration inside the cell is greater than the outside.

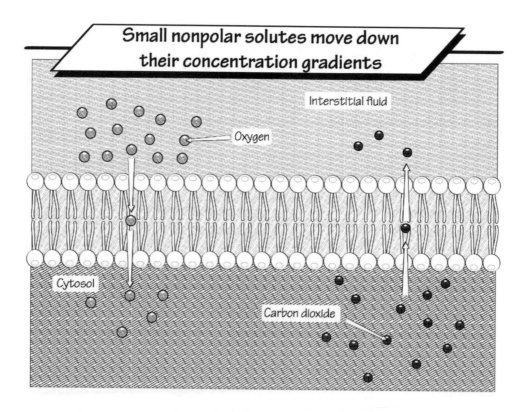

Diffusion requires no energy because it's a downhill motion. Like rocks naturally falling down a hillside from a high energy state on top to a low energy state on the bottom, gases and hydrophobic molecules go downhill, from where the concentration is so high that the molecules are bumping into one another to a low energy state where they are not.

Again, diffusion across the cell membrane is possible for some molecules but not for others. With regard to your cells, there aren't too many molecules that can truly benefit from the process of diffusion across a membrane.

Facilitated Diffusion

Facilitated diffusion is basically diffusion with help. It is still diffusion from an area of high concentration to an area of low concentration, and it still requires no energy to do this. You can think of facilitated diffusion as the cell membrane's way of getting past its hydrophobic nature, to allow polar molecules to benefit despite their hydrophilic nature.

Facilitated diffusion is the passive transport of molecules (including ions) across the cell membrane, with the help of specific proteins in the cell wall. There are two types of proteins embedded in the cell membrane that help these polar molecules diffuse across it. These are channel proteins and carrier proteins. The next figure shows what a channel protein and carrier protein look like:

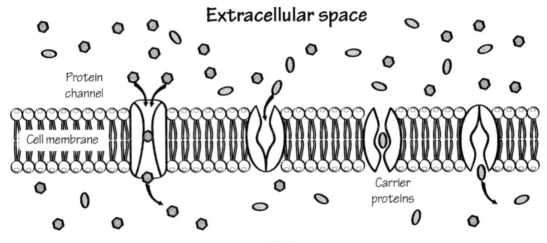

Facilitated diffusion is different from plain diffusion in many ways. While it is still diffusion and a downhill process from high concentration to low concentration, the transport in facilitated diffusion needs a protein that has a hydrophilic center to send the molecule through the cell membrane. The polar molecule no longer has the hydrophobic barrier in the phospholipid layer that prevents its passage through the cell membrane.

Another way that facilitated diffusion is not the same as free diffusion is that it is not linear. Regular diffusion happens all over the cell membrane, and the gradient is always from high concentrations of a substance to low concentrations in a straight line between concentration levels. Facilitated diffusion instead depends on how many of these channel or carrier proteins there are, and how quickly they can do their job. This means that there isn't a smooth transition as the molecules cross the membrane, like there is with simple diffusion.

It would be like transporting a bunch of people via taxi. You can only do it if there are enough taxis to transport them. If there are a lot of taxis for every person, the process goes quickly. If there are just a few, some have to wait while others cross over using the number available. In the cell membrane, lots of channel and carrier proteins means diffusion is rapid. If there aren't too many, the process does not go as smoothly or as quickly.

Chloride, sodium ions, potassium ions and glucose are just a couple of examples of molecules and ions that need to cross the cell membrane but cannot do it by simple diffusion, because they're hydrophilic. They need to pass by means of facilitated diffusion to avoid being affected by the lipid layer in the cell membrane.

Let's look first at how channel proteins help in facilitated diffusion. Channel proteins are just that: They are channels through which ions and small molecules pass. If they are open all the time, they are called *non-gated channel proteins*. Their main job is to balance the levels of water and ions on both sides of the cell membrane. One channel protein is called *aquaporin*, a channel protein that forms pores in the membrane of biological cells. Its main purpose is to facilitate transport of water between cells.

Gated channel proteins stay closed unless there is a signal to tell them to open. The signal can be a chemical one or an electrical one. Gated channel proteins are nice because they help the cell have a negative charge on one side and a positive charge on the other side. This is super important in nerve cells and muscle cells. You see now how this is still diffusion but with a twist. Gated channel proteins do allow molecules to cross over from high to low concentrations, but only when the circumstances are right.

Here's how it works in muscle cells. A muscle cell has a place called the *neuromuscular junction*. When activated at the junction by a nerve impulse, a chemical signal goes from the tip of the nerve cell to the muscle cell. This flips a switch so that channel proteins can open up and allow sodium to flood into the cell. Sodium, being positively charged, causes more electrical positivity in the muscle cell. The muscle will respond to this by contracting, although later you will see it is a more complex process than this. There are other mechanisms we will soon talk about that return the cell to its resting state. This figure shows how it happens:

Carrier proteins are another way of allowing facilitated diffusion to take place. These are not the same as channel proteins, because they need whatever is trying to cross the membrane to bind to the protein first. This is dependent on how much of the molecule there is, how many carrier proteins there are, and temperature. Temperature is important because it raises the kinetic energy, so that individual molecules become more jittery and have a greater chance of bumping into a carrier protein.

Carrier proteins work best with small molecules like glucose that are too big to fit through channel proteins. There are some channel proteins called *uniporters* that bind a single glucose molecule, changing their shape or conformation in order to squeeze the molecule through the membrane, spitting it out on the other side.

There still is no energy necessary to carry out this kind of facilitated diffusion. Carrier proteins are just built to bind, morph into a different shape that draws the molecule in, and then change shape as they spit the molecule out on the other side. By itself, carrier proteins that do this cannot go against the concentration gradient across the membrane. The direction is always from a high concentration to a low concentration.

One difference between channel proteins and carrier proteins is the way they decide to let a molecule get through. Channel proteins don't actually bind to the molecule that passes through them, but they are picky about things like the size and electrical charge of the passing molecule. Carrier proteins, on the other hand, have a specific binding site for the molecule they help get through the membrane. If the molecule can't bind, it won't pass through. This figure shows what a carrier protein does when it encounters a glucose molecule:

Facilitated Diffusion

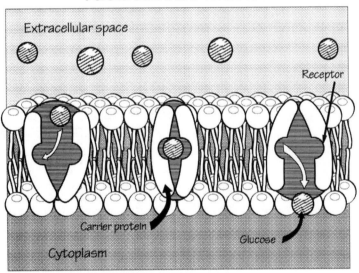

Active Transport Needs Energy!

Active transport across a cell membrane also involves a carrier protein. It still needs something to bind to it before it can do its job. There's a big twist here, however. When a carrier protein uses active transport, it carries molecules in the opposite direction of its concentration gradient. This requires energy.

Think of it like lugging a pail of rocks up a hill. Does it take more energy to carry the pail up the hill or down? Anyone who's done this will tell you it takes much more energy to go uphill than down. The rocks and the pail are the same, but the energy to move them depends a lot on the terrain you pass through.

Active transport goes against the natural tendency for substances to pass through a membrane from a higher to a lower concentration. While passive transport uses kinetic energy in order for it to happen, active transport relies on energy made by the cell.

Active transport helps a cell build up high concentrations of substances it needs from outside the cell, such as amino acids, glucose and enzymes. It is called *primary active transport* if adenosine triphosphate (ATP) is used to power the ion channel. It is called *secondary active transport* if it uses an electrochemical gradient like the difference in sodium concentration across the cell membrane to pass the needed substances through the membrane.

ATP is Your Energy Source for Active Transport

ATP is a nucleoside triphosphate, which is full of energy because it has three phosphate molecules attached to adenosine. These three attached phosphate molecules are loaded with chemical energy. When one of these is removed to make ADP, or *adenosine diphosphate*, energy is given off and put to work to help molecules cross the membrane in an uphill way.

Active transport using ATP is a lot like having a conveyor belt driven by electrical power that helps get the pail of rocks up the hillside. Without the electrical energy applied to the conveyor belt, it's useless. In the same way, without ATP energy, the carrier protein used to help molecules pass across the membrane would be useless too. The following figure shows what ATP looks like, and how it turns into ADP in order to release energy. Notice the three phosphate groups attached to it.

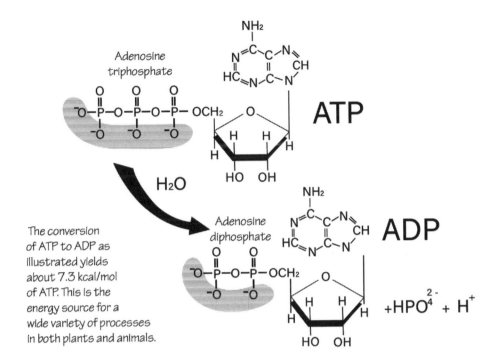

The conversion of ATP to ADP as illustrated yields about 7.3 kcal/mol of ATP. This is the energy source for a wide variety of processes in both plants and animals.

The metabolic processes that use ATP as a form of energy convert it back to ADP, which gets recycled in the mitochondria to make more ATP energy. With three phosphates, ATP is a high-energy molecule. When it loses a phosphate group to become ADP, it becomes a low-energy molecule. During the chemical reaction that turns ATP into ADP, this energy difference is used as an energy source to drive other reactions.

ATP is used in a variety of cellular processes besides active transport, such as the production of DNA, RNA, and many types of proteins. A great portion of it, though, is used for cell membrane active transport. In fact, about 20 percent of all your ATP is used in membrane transport, particularly at the sodium-potassium ATPase pump.

Fun Factoid: *When you go for a run, your muscles use glucose for energy. In order to get glucose into the muscle cells, you must have ATP energy. For the first 8 to 10 seconds of your run, the muscles cells can quickly make ATP out of a high-energy molecule called* creatine phosphate. *In order to go for another 90 seconds, your muscle cells break down its stored glucose in the cell, called* muscle glycogen, *to make ATP for your muscles to work. After that, hopefully your circulation is flooding your cells with oxygen. Your mitochondria undergo* aerobic respiration *using this oxygen and glucose from outside the cell in order to make ATP energy to go the distance. Without this third source of ATP energy, you would burn out after just 90 seconds.*

The Sodium-Potassium Exchange Pump

Sodium and potassium ions need to travel from one side of the cell membrane to the other against their natural concentration gradient. Because they do not tolerate the hydrophobic environment of the cell membrane *and* they travel from a lower concentration to a higher concentration, they need a pump in order to transfer them from one area to another. Enter the sodium-potassium pump, which is also called the *sodium-potassium ATPase pump* because it needs ATP energy.

This important pump is responsible for transferring sodium from the inside of the cell to the outside, and for transferring the potassium from outside the cell to inside against their concentration gradients. It's what keeps a higher concentration of sodium outside the cell and a lower concentration of potassium inside the cell. When the pump works the way it's supposed to, each time it's activated it sends three molecules of sodium outside the cell for every two molecules of potassium ions inside the cell.

When the pump works, it uses an enzyme called *ATPase* sitting just next to the membrane, that takes ATP and turns it into ADP, releasing energy. At the same time, it binds three sodium molecules in the cell and two potassium molecules outside the cell. ATPase acts like a battery whenever ADP is made from ATP. This energy goes into pushing the sodium molecules out and the potassium molecules in. This figure shows what the pump looks like:

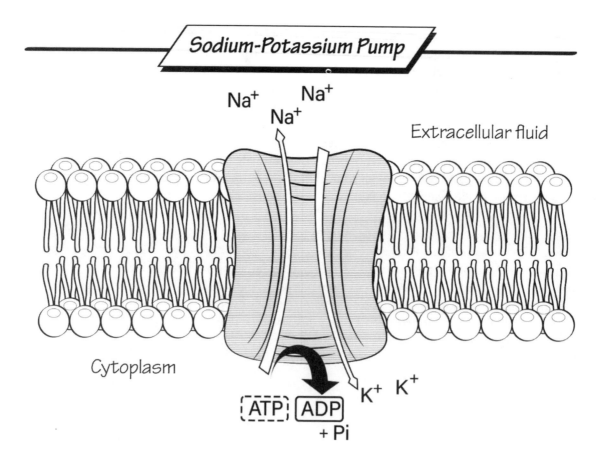

If you look carefully, you can see how this pump causes the inside of the cell to have a greater negative charge than the outside. If two positive potassium molecules enter the cell for every three positive sodium molecules, there is a charge imbalance going on. Because this pump is running all the time, the effect will be more positive charges outside the cell than inside. It means the inside of the cell is almost always less positive, or more negative, compared to the outside.

Uniport, Symport, and Antiport Mechanisms

We've already talked briefly about uniport transport mechanisms in cell membrane transport. The term uniport means "one direction." This involves a molecule that goes in one direction only, through the use of a carrier protein. It can happen in facilitated diffusion and it can happen in active transport. But you should know about two other processes that fall under the umbrella term of secondary active transport. These are more complicated than the sodium-potassium pump, because there are more things to do to get a molecule to pass through the membrane against its concentration, or energy gradient.

Secondary active transport requires another ion for it to work. This ion is called a *cotransporter*. Sodium ions are common cotransporter molecules. When this type of transport is used, sodium passes through the carrier protein along its concentration gradient, from high concentration levels to low concentration levels. This powers the part of the carrier protein that can drive another molecule to also pass through the carrier protein against its concentration gradient, or "uphill."

In a sense, secondary active transport takes the energy of a process that occurs naturally (like sodium crossover along its concentration gradient) to drive a process that wouldn't occur naturally (like glucose crossover against its concentration gradient).

Symporter secondary transport involves two molecules going through the carrier protein in the same direction as one another. There is a symporter mechanism in intestinal cells lining the small intestine called SGLT1. It helps draw in glucose from the gut to the intestinal cell. Sodium crosses into the cell along its natural concentration gradient, and takes a glucose molecule along with it in the same direction, but against a concentration gradient. This figure shows how it works. Locate the number 2 within the illustration to see the difference between secondary transport and primary active transport.

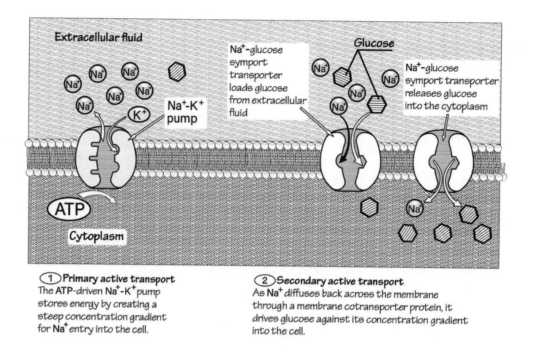

① **Primary active transport**
The ATP-driven Na⁺-K⁺ pump stores energy by creating a steep concentration gradient for Na⁺ entry into the cell.

② **Secondary active transport**
As Na⁺ diffuses back across the membrane through a membrane cotransporter protein, it drives glucose against its concentration gradient into the cell.

An *antiport transport mechanism* is similar to the symport mechanism, because it still requires sodium to cross the membrane along its gradient; but instead of drawing in another molecule along with it in the carrier protein, it draws another molecule against its concentration gradient in the opposite direction of that traveled by the sodium ions. The following figure shows how this works for the transfer of an amino acid molecule across the membrane. There is also a sodium-calcium exchanger pump that does the same thing.

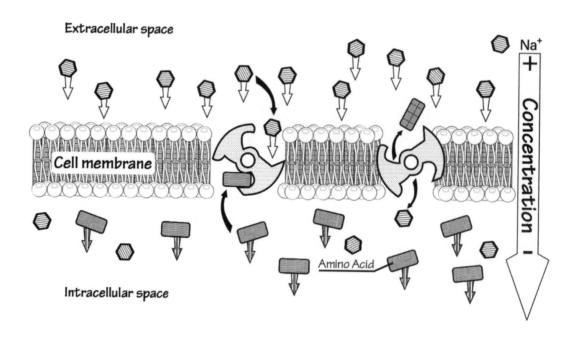

Uniport transport is driven by ATPase and the transfer of ATP to ADP, giving off the energy needed for the molecule to pass through the membrane. Symport and antiport mechanisms rely instead on the electrochemical differences between the inside and outside of the cell to drive the transport of molecules needing energy to pass through the cell for any reason.

Endocytosis

Endocytosis is a completely different type of active transport than any we have already talked about. With endocytosis, a cell transports proteins and other molecules into the cell using *vesicles*. Vesicles are small, fluid-filled bubbles that can carry large or small molecules into the cell in numbers that are larger than can happen in other kinds of active transport. They cannot transport molecules through the cell membrane by themselves, and need to be engulfed by the cell, taking pieces of the cell membrane to make the vesicle. The vesicle enters the cell and travels to wherever it is needed inside the cell. Endocytosis involves two similar processes called cell eating (also called *phagocytosis*) and cell drinking (also known as *pinocytosis*).

Exocytosis is endocytosis in reverse, where small bubbles or vesicles are made inside the cell and sent in large groups of molecules outside the cell. The contents of a vesicle leaving the cell can be waste products or molecules the cell wants to send out as messengers to other cells. This figure shows what endocytosis and exocytosis look like:

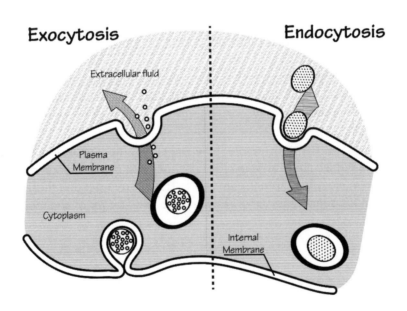

Endocytosis doesn't happen randomly. The cell needs some kind of signal to tell it to do this. There are four separate methods. There are proteins called *clathrins* that collect on one side of the cell membrane. They collect themselves into a cage that sucks in the membrane and the molecules in them.

Receptor-stimulated endocytosis relies on a receptor on the cell that gets stimulated by a chemical messenger. Once this happens, the same clathrin proteins get together and start the endocytosis process in exactly the same way as it does without a receptor. The end result is the ingestion of molecules into the cell. This figure shows how clathrin sucks stuff up into the cell during endocytosis:

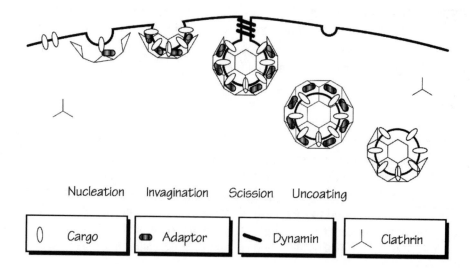

Nucleation   Invagination   Scission   Uncoating

| 0 Cargo | ● Adaptor | ＼ Dynamin | ⊥ Clathrin |

There's one other way that endocytosis can happen that doesn't need the clathrin proteins. This process uses caves called *caveolae*. They are made from an integral membrane protein called *caveolin*. When these proteins get activated, they make caves, or collection pits, that collect molecules in order to bring them inside the cell.

To Sum Things Up

The cell membrane is probably the most important part of the cell. It determines what gets into the cell, what leaves the cell, and how much of each get from one side to the other. As you have seen, there are many processes that can accomplish this. You've learned about osmosis, which involves the transport of water from one side of the cell membrane to another. The other processes of diffusion and facilitated diffusion also do this, and without expending any energy.

Active transport usually means the transfer of a molecule across the membrane, from an area of low concentration to an area of high concentration. This cannot happen unless some type of energy is involved. Endocytosis and exocytosis both use energy to transport larger and more diverse molecules across the membrane at the same time, using vesicles instead of carrier or channel proteins.

Chapter Five: DNA, RNA, and Cell Reproduction

It's hard to understand the whole of the anatomy and physiology of the human body without digging deep into the core of what makes cells divide, grow, and do the things they need to do in order to function as a cell, tissue, or part of an organ. In this chapter, we will look at the building blocks of the cell, which are its *nucleic acids*. We will look at how they participate in cell reproduction and how nucleic acids get turned into the proteins that are the workhorses of the cell.

The cell's nucleic acids and their interaction are a lot like the nervous system of the body. The DNA of the cell is its brain, which provides all the information necessary to tell the cell what to do. The RNA and ribosomes of the cell are the wiring, or peripheral nervous system, of the cell. They take the DNA message and, through several steps, send the original message through pathways that lead to the making of proteins. Proteins then do their job to manage biochemical pathways as enzymes, or to make structural proteins used to build the cell components.

All of this works in a kind of chemical dance that allows several "dancers" to start the message in the DNA, send it down the line, and spit the message out as a protein molecule that does all the real back-breaking work needed to operate the cell on a continual basis. Let's look first at the brains behind the operation, which is the cell's DNA.

DNA and RNA

DNA lives inside the nucleus of the cell and is responsible for sending chemical messages to all parts. It contains the blueprints for the synthesis of proteins and other molecules. RNA takes the message from the DNA and goes on to the ribosomes, where it translates the message into proteins used by the cell. This figure shows what the DNA double helix looks like:

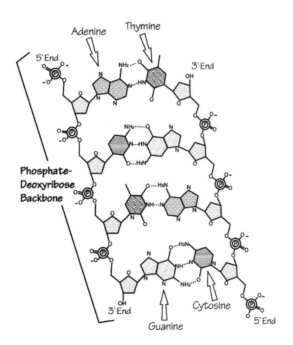

Complementary Base Pairing

DNA stands for *deoxyribonucleic acid*. It is a *polymer*, which means it's a chain of separate units called *monomers*. The monomers in the case of DNA are called *nucleotides*, which are a mixture of an *organic base molecule*, a *deoxyribose sugar*, and a phosphate group. This figure shows you what a nucleotide looks like:

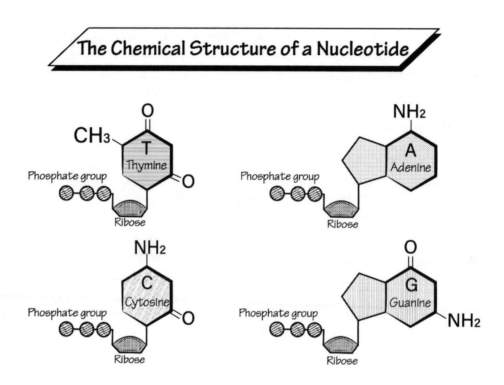

DNA is made from a long line of four separate types of nucleotide bases, each of which pairs with only one of the other bases. Nucleotide bases are unique molecules that, when lined up in different ways on the DNA molecule, tell a certain story about what the cell is all about and what it needs to do. The four bases used in making DNA include *guanine, cytosine, thymine,* and *adenine*. The cytosine base always pairs with the guanine base, while the thymine base always pairs with the adenine base.

A DNA molecule is a long string of these bases that are in a double helix formation. It is comprised of two strands that look like a ladder. Each rung of the ladder is a pair of connected bases called A-T or G-C base pairs. The DNA strand is not a naked ladder but is mixed in with proteins that help it form *chromatin*. One line of chromatin is called a chromosome. Since every base pairs with only one type of other base, one can determine the DNA sequence by knowing the base pairs of a single strand.

The picture below shows what chromatin looks like. There are proteins called *histones* that form *nucleosomes* scattered along the DNA strand like beads on a string:

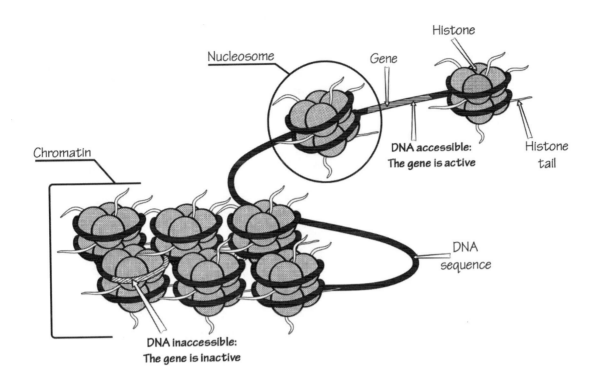

Histone proteins are important because they coil up DNA so that it isn't so long. They also represent areas of the DNA that are blocked from being read by RNA molecules. This is just one way that your body's cells know which genes to read and which to ignore. Those that are tightly bound into nucleosomes by histone proteins just aren't accessible to DNA-reading enzymes and can't be transcribed, which would otherwise start the process of making a protein.

The two sets of complementary base pairs are commonly represented in a form that is abbreviated. For example, the cytosine-guanine base pair is also called C-G, while the thymine-adenosine pair is also called T-A. Each base pair is connected along the strand by a couple of *hydrogen bonds*. The C-G pair is linked by three hydrogen bonds, while the T-A pair is linked by two hydrogen bonds.

DNA has an interesting structure in that the two strands run opposite to one another. One strand is called the *5' to 3' strand*, while its matching strand runs in the opposite direction, called the *3' to 5' strand*. Because DNA is made of strands that are not the same on both ends, and because the strands connect opposite to one another, these are called *antiparallel strands*.

RNA, or Ribonucleic Acid

RNA, or ribonucleic acid, is very similar to DNA because, like its DNA cousin, it is also a polymer or string of nucleotides. There are some differences between the two molecules you should know about. RNA can be double-stranded but it is very rarely found this way. Instead, it is single-stranded but can form base pair matches with itself so that it isn't linear, but rather made of segments of hairpin loops where the strand has folded back on itself. It also has ribose as its sugar on the nucleotide instead of deoxyribose. Finally, instead of the same A, G, T, C base pairs of DNA, thymine is replaced with *uracil*, or "U," in the RNA molecule.

RNA comes in different types that have the same basic structure. There is messenger RNA, or mRNA. These are long strands that take the DNA message, leave the cell nucleus, and join with ribosomes to make proteins. There is also transfer RNA, or tRNA, that is shorter the mRNA. It forms a unique shape that allows it to be charged with an amino acid so it can help to make a protein strand in the ribosome. There is also ribosomal RNA, or rRNA, that makes up a large proportion of the ribosome itself. Finally, small nuclear RNA, or snRNA, does many jobs, including helping messenger RNA to be readied in the nucleus before it can send on the DNA message.

The following figure shows what transfer RNA looks like. Notice the hairpin loops that make up its shape:

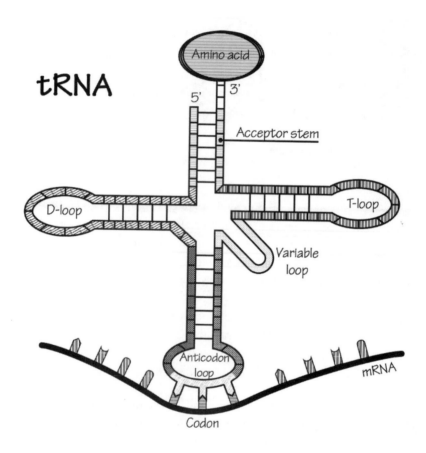

DNA Replication

Replication is how DNA makes copies of itself when a cell is trying to divide. In making these copies, two of the same pieces of DNA are made. One copy goes to one of the daughter cells after the cell has divided, while the other goes to the other daughter cell.

We refer to DNA replication as *semi-conservative replication* for one main reason. In order to make a copy of itself, the DNA strand splits in certain places and new nucleotides attach to each of the split strands. When the process is finished, there are two separate strands. Each double strand has one side made from the original parent DNA strand and one brand new copy. One strand is conserved, or recycled, from the old double strand of DNA, and one strand is created using the old strand as a template.

Because A only binds with T and G only binds with C, each strand is a blueprint telling an enzyme called *DNA polymerase* to create a matching or complementary opposite strand. DNA polymerase is a complicated enzyme that can not only make DNA strands but can also proofread itself to make sure it didn't make any mistakes.

There are other important enzymes that help DNA polymerase replicate DNA. There is *helicase*, which unwinds and separates the strands. There is also *topoisomerase* that keeps the single strands steady so they don't get tangled in the DNA replication process.

There is also DNA *primase* that makes a short RNA strand segment on the strand being replicated. This is necessary because DNA polymerase cannot just start making DNA anywhere on the strand. This short piece of RNA made by DNA primase helps DNA polymerase know where to start. The short pieces of RNA made in the process get cut out and replaced with DNA strands later.

Finally, there is *DNA ligase*, which attaches individual DNA strands so that they form one continuous strand. This figure shows what the process would look like if you could see it happening:

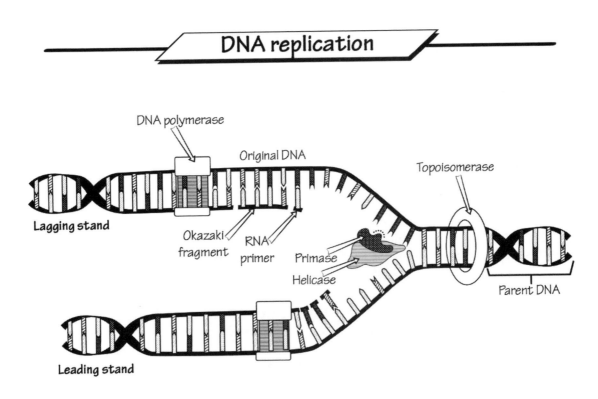

So, how does DNA replication work? It's actually kind of complex. There is the *initiation phase*, where DNA helicase separates the double strand of DNA along the chromosome. This creates a V-shaped part called the *replication fork*.

One downside of DNA polymerase is that it can only make new DNA pieces from the 5' to 3' direction. When it makes DNA on one strand, this is okay because the strand runs in that direction, starting at the replication fork. This is called the *leading strand*.

The other strand isn't as lucky, so it's called the *lagging strand*. DNA polymerase can't just start at the replication fork because it would have to make DNA in a backwards direction. To accommodate this, RNA primase makes short RNA segments and short pieces of DNA called *Okazaki fragments*. Those get made in the correct direction (5' to 3'). When all of the fragments have been made, the RNA segments are cut out and DNA ligase connects the Okazaki strands together. Voila! You have a perfect backward strand.

The termination of DNA replication happens when the two strands get copied and when all the replication forks are finished with the replication process. When it works perfectly, there are now two separate strands of DNA that are identical to one another. The process isn't perfect, but with proofreading, only one in a billion nucleotides are incorrectly placed. Without proofreading, though, there would be about 120,000 mistakes per cell with each cell division.

Fun Factoid: *We are all mutants! DNA replication is nearly perfect but mistakes happen all the time. Fortunately, the mutant cells made by these mistakes won't kill you most of the time. Some of these daughter cells can't survive. Others have a mutation that makes no difference because it happens at a spot in the DNA molecule where no major proteins get made. A few have mutations in parts of the DNA with a gene that regulates cell division itself. This is bad because it can lead to uncontrolled cell growth or cancerous cells. Fortunately, your immune system is on the lookout for these damaged cells and kills them on a daily basis.*

Transcription

DNA transcription is similar to replication except that, instead of making a second DNA strand, it makes a complementary RNA strand. The purpose of transcription is different from replication. While replication is used to make chromosomes for cell division, transcription happens all the time in order to make messenger RNA that will leave the nucleus to make proteins.

You can think of DNA as the post office and messenger RNA as the letter to be delivered to a ribosome residence. In this case, the post office creates the letter, hopefully without too many mistakes. The messenger RNA letter goes to the ribosome to deliver the message. The message is read by the ribosome and a protein is made based on the message.

In transcription, the main enzyme used is called *RNA polymerase*. Unlike its DNA polymerase cousin, RNA polymerase does not need a starter template, but knows exactly where to start transcribing the DNA message. It makes much shorter RNA strands than DNA polymerase, which makes an entire chromosome at one time. RNA polymerase makes more errors but these are not as disastrous because they don't affect the whole cell's DNA message for the rest of time, as is true of DNA replication errors.

The main similarities between replication and transcription include that the process still involves making a complementary strand of a nucleic acid. The RNA polymerase molecule can also only operate in one direction, and makes RNA in the 5' to 3' direction. Fortunately, there is no problem with this because just one strand gets transcribed, and it happens to be the one that goes in the right direction. This figure is what RNA transcription looks like:

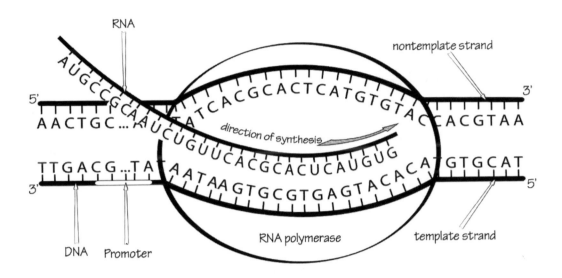

There are different types of RNA polymerase inside your cells' nuclei that make the different types of RNA. RNA polymerase II makes only messenger RNA. RNA polymerase I makes only ribosomal RNA. RNA polymerase III makes only transfer RNA. When messenger RNA is made, it isn't finished getting processed, so it is called pre-mRNA.

When pre-mRNA gets made, it is too long and contains a lot of nucleotides it doesn't really need. It becomes the job of a protein complex called a *spliceosome* to cut out the unnecessary junk RNA. The segments of RNA you need to make a protein are called *exons*, while the parts cut out are called *introns*.

Because naked messenger RNA can get attacked and damaged on its journey to the ribosomes, it is capped on one end (called a *5-prime cap*) and has a tail of adenine-based nucleotides attached to the other end (called a *poly-A tail*). These help to protect the molecule so it can get from the inside of the nucleus to the outside safely.

Translation

Translation happens at the ribosomes. By now, the DNA at the nuclear post office has sent the messenger RNA letter to the ribosomes. Now that the message is actually in the ribosomes, this message must get translated into a language the cell can understand. The original message is written by a series of nucleotides on the DNA and RNA molecules, but the final translation must be in amino acid language—the language of proteins used by the cell.

In the ribosomes, the messenger RNA molecule attaches like a long string of letters. The machinery of the ribosomes gets activated and the message is read. How is this message interpreted when it is just a string of nucleotides? It turns out that there is a secret code embedded into our cells that is shared by almost all life forms. This is called the *genetic code*. It involves combining three nucleotides at a time into what's called a *codon*. Each codon tells the ribosome which amino acid it needs to attach.

If there are four different nucleotides and each codon is a cluster of three of them, this leaves a total of 64 different possibilities of codons. As there are just 20 possible amino acids in the protein molecule, this number is plenty to have at least one codon matching a unique amino acid. In fact, there are some amino acids that have two different codons that code for them. There are also a few codons that say "start translation" and a few codons that say "stop translation." These codons are used to tell the ribosome where to start making a *polypeptide chain* and where to stop doing it.

This figure shows you what translation in the ribosome looks like:

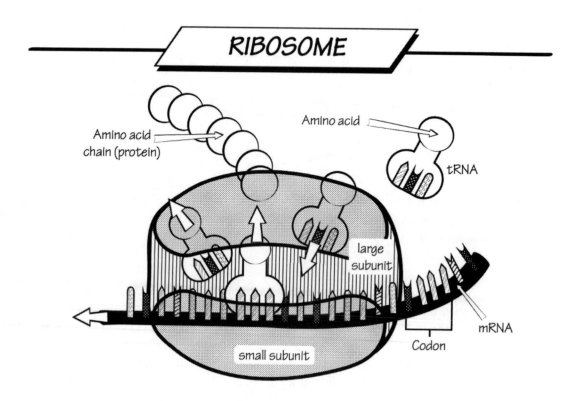

You can see here that there is the messenger RNA molecule with its codons, or clusters of three nucleotides. It comes into contact with the transfer RNA molecules—each of them loaded with an amino acid. There is a segment on the tRNA molecule called an *anticodon* that matches with the codon on the mRNA molecule. If there is a match, the right amino acid is attached to the growing polypeptide chain. One-by-one, the polypeptide chain grows until a stop codon is reached that detaches the polypeptide chain.

Ribosomes can be free in the cytoplasm, but most of the time they are attached to the endoplasmic reticulum. The proteins made by the ribosomes get dumped into the endoplasmic reticulum, where they get further modified. There are slight changes made to some amino acids, and some areas of the polypeptide chain that are deleted, to make a final three-dimensional protein shape.

There is also a proofreading mechanism in the endoplasmic reticulum, so if the protein is nonsensical or wrong in some way, it gets chewed up and recycled at a later time. Proteins that are normal and functional often get sent onto the Golgi apparatus, which is another type of post office or dispatcher that sends each protein molecule to the place in the cell where it is most needed.

Cell Reproduction

There are two ways a cell might reproduce itself. The simple way is to make a copy of its *genome*, or set of chromosomes, divide these into two equal piles, separate them, and then make a division between these to create two new identical daughter cells. This is called *mitosis*. Most cells can do this; it's how new cells get made all over your body every single day.

You have 46 chromosomes within each of the nuclei in your cells. In mitosis, they get replicated in the way we just talked about, so that temporarily there are double that number of chromosomes in the cell. As mitosis happens in an orderly fashion, the duplicated chromosomes, called *sister chromatids*, get separated. In this figure, you can see how mitosis is supposed to work:

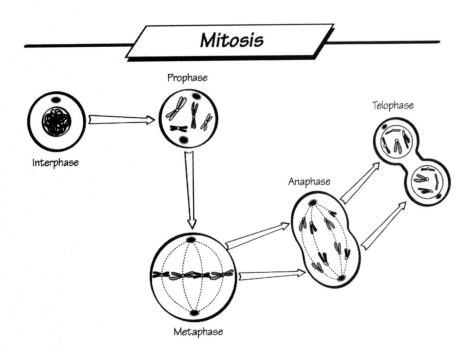

There are several phases to this process. Most of the time, the cell is in *interphase*, where it is business as usual. Proteins get made and the cell grows without any major visible events happening. If a cell is going to divide, however, DNA gets *synthesized*, or replicated, near the end of interphase in preparation for mitosis to occur.

Mitosis is cool because you can see it happen under the microscope. Normally you cannot see DNA under the microscope, but in mitosis the chromosomes thicken and condense so you can see them clearly under a regular light microscope.

In *prophase*, the chromosomes are visible but still look tangled. As *metaphase* approaches, the chromosomes, with the connected sister chromatids attached to each other, get lined up along an invisible metaphase plate. Then the microtubule machinery in the cell gets into gear and drives the sister chromatids apart in *anaphase*. They are drawn to opposite sides of the cell and are separated in *telophase*. Finally, a process called *cytokinesis* occurs, which is basically when there is a pinching off of the cell membrane between the two clusters of chromosomes to make two separate daughter cells.

Mitosis occurs only in *eukaryotic* cells like animals, plants and fungi. It is different than the way *prokaryotes* like bacteria divide, even though the end result of two identical daughter cells is the same for both types of organisms. Prokaryotic cells have no nucleus, so they divide by a different process, known as *binary fission*.

Fun Factoid: *Walther Flemming was a cytologist in the late 1800s who first noticed mitosis going on, except that he didn't know what he was seeing. Finally, he noticed the same thing happening regularly in salamander cells and described the whole process—even naming all of the different parts of the process we still use today. He finally published what he knew about mitosis in 1882, paving the way for greater understanding of how cells are able to divide and grow.*

Meiosis

Meiosis is a special type of cell division that decreases the number of chromosomes in the parent cell by one half. It only happens in human *germ cells*, such as those found in the female ovary and the male testes. These germ cells divide in a more complex way in order to form unique *haploid gametes*. Gametes are sex cells—either a sperm cell or an egg cell—that participate in fertilization in order to create unique individual offspring.

Your normal cells are called *diploid cells* because there are two pairs of similar chromosomes in each cell. The chromosomes are called *homologous chromosomes* because they are similar but not identical to one another. In other words, when you were first created, you got one chromosome of each type from each of your parents. The chromosomes you got from your dad weren't the same as from your mom. How could they be if your dad had blue eyes, for example, and your mom had brown eyes? They each gave eye color genes, but they were homologous—not identical.

Your parents' germ cells divided to form four haploid gametes. Haploid means that, instead of having 46 chromosomes, each gamete has 23 chromosomes. The male gamete is called a *spermatid*, or sperm cell, while the female gamete is called an *oocyte*, or egg cell.

During reproduction, the sperm and egg combine to make a zygote—a one-celled version of a human being yet to develop. This zygote is diploid and has 46 chromosomes. Zygotes then divide through the process of mitosis in order to have rapid expansion of cells in just a few days after fertilization.

Meiosis is different in more ways from mitosis than the fact that it makes haploid cells. These haploid gametes could not possibly be as identical to one another as the chromosomes that are made through mitosis. If this were the case, there would not be such variations in how you and your siblings look from one another. You would have too great a chance of looking exactly like your brother born two years after you.

Meiosis is ingenious. Because it is different from mitosis in one unique way, the gametes created are all different from one another. This all happens because of what's called *recombination*. Early in meiosis, the germ cells line up the homologous chromosomes during the first prophase part of the process. These chromosomes then mix and match segments of DNA with each other in a kind of musical chairs so that every chromosome is then rearranged in a new way. When they finally separate at the end of meiosis, the chromosomes in each gamete are not identical in any way to the chromosomes they started out to be.

Fun Factoid: *Meiosis is not a perfect process. If the recombination process goes poorly, one chromosome can get too many of the same gene, while others don't get the gene at all. Those that don't get the gene at all often die or create a baby with some type of birth defect. If the actual chromosome separation process fails in what's called* nondisjunction, *one gamete gets an extra whole chromosome, while the other gets none. Diseases like Down syndrome, or trisomy 21, happen when a gamete gets two copies of a chromosome. In Down syndrome, one gamete gets two chromosome 21 copies and mixes with one copy from the other parent to make three copies in the zygote. Almost all parts of the body are affected by this major birth defect.*

The following figure shows how meiosis takes place. As you can see, there are two main phases, called meiosis I and meiosis II. When both are complete, you have four gametes out of one original germ cell. Because of recombination in meiosis I, these gametes are all original and different from one another.

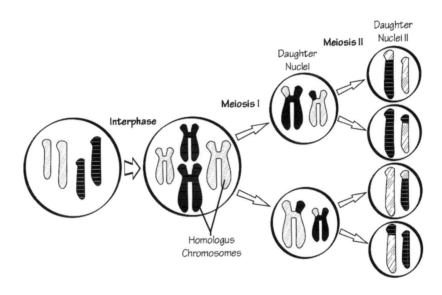

There is one difference between the way female germ cells make gametes and the way male germ cells make gametes. When males do this process, it creates four unique sperm cells, each of which has very little cytoplasm, is very small in size, and is motile. They have to be motile in order to travel to the female egg, which is just sitting there waiting to be fertilized.

Females do this differently. The process of meiosis is essentially the same but the end result is not four small haploid egg cells but just one huge haploid egg cell. What happens to the other three? These get made but the process isn't symmetric. One daughter cell gets all the cytoplasm during meiosis I of cell division, while the other gets very little. It turns into an inactive polar body that simply degenerates. This happens again during meiosis II, so that there is a polar body and a normal egg cell.

If you do your math properly, you will see that one germ cell gives rise to just one haploid egg cell and three polar bodies. This egg cell is unique just like sperm cells are, but there is just one of them left alive and full of cytoplasm, ready for fertilization. This figure shows the fertilization process:

Reproduction Process of Human

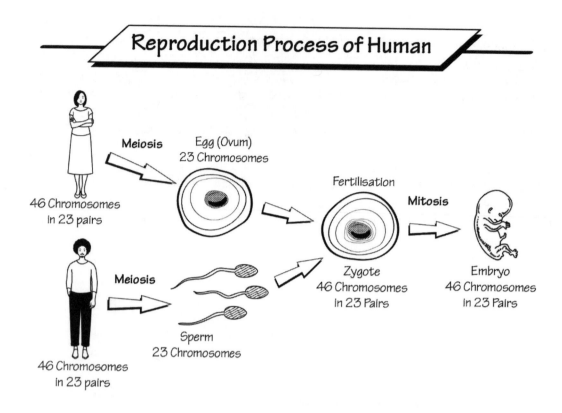

When sperm and egg come together, the male gamete provides its DNA in the form of 23 chromosomes, and the egg cell provides its own 23 chromosomes of DNA (plus almost all the cytoplasm necessary to make a new human being). Something mysterious then happens and a new person is made in the form of one diploid zygote. This zygote soon begins dividing rapidly through mitosis to start the vast developmental process needed to make an embryo, then a fetus, then a baby, and finally—you. This is just a snapshot of what the amazing process of creating new life is all about.

Fun Factoid: Are you more like your mother or your father? Regardless of what you actually look like, it is safe to say that, genetically speaking, you are more your mother's child. This is because, while equal numbers of nuclear DNA come from each of your parents, the same can't be said of the DNA that exists in the mitochondria you inherit, called mitochondrial DNA. *Your mitochondria makes its own DNA in small amounts and, because the egg cell is huge and has all the cytoplasm, it also donates all of the mitochondria (and its DNA). The sperm cell rarely, if ever, contributes to this. This makes you genetically more like your mother. Ancestral geneticists have used this fact to track the maternal line of every person on Earth back to the early years of mankind.*

To Sum Things Up

The process of making new cells involves the replication of DNA and the separation of the chromosomes in the process of mitosis. DNA is where the cell's brain is located. It contains all of the genetic material your cells need to make all of the proteins and other substances necessary to operate the cell. There are mechanisms that make sure each cell, while having your entire genetic material, knows which proteins it does and doesn't need to make.

The process of making proteins requires a similar process to replication of DNA, except that it involves an enzyme complex called RNA polymerase that makes RNA instead. One type of RNA, called messenger RNA, delivers the DNA message to the ribosome. Through the reading of the genetic code, the nucleotide message gets translated into amino acids and proteins in the ribosome. These proteins get further processed in endoplasmic reticulum before doing their various jobs. None of these processes directly make carbohydrates or lipids—only the protein enzymes necessary to make these other molecules.

Cells can divide through mitosis or meiosis. In mitosis, two identical diploid daughter cells get made. This happens all the time all over the body. Meiosis, on the other hand, happens only in the body's germ cells in the ovaries or testes. The goal is to create haploid gamete cells (sperm or egg cells) that get fertilized in the female to create a diploid zygote cell—the starting cell of every human being.

Chapter Six: Your Skeleton Explained

In this chapter, we will talk about you in your most structural sense, as we discuss your bones. While we talk about your skeleton and all of its many parts, we will also review the amazing properties of the bones and joints themselves. Bones are not just stiff and dead parts of you that hold up your insides. They are dynamic, which means that they have their own metabolism. Bone cells and minerals are added and removed all the time. So much of the health of your bones depends on the vitamins and nutrients you eat every day, as you will soon learn. At the end of this chapter, you might not think about your bones the same way again.

What is the Skeleton?

Your bones all together make up your skeleton. It's just like the framework of your house. Without the stud walls and two-by-fours inside the walls, nothing inside would be protected from the weather, and it would be a fragile structure to live in. The difference between the framework of your house and the framework of you is that your skeleton is not just built and forgotten. It changes all the time and is responsive to the everyday stresses on it.

You have almost 300 bones when you're born but only about 206 bones as an adult. No, you don't actually lose any bones, but instead some infant bones will fuse together at different times in childhood and adolescence, which shrinks the total bone count over time.

You've probably heard of *osteoporosis*, where your bones are too thin and prone to breakage. While this is a disease mostly seen in older people, you actually start losing your own bone mass starting at the age of twenty, which is when your bone mass and density is the greatest. It all goes downhill from there, even if you eat healthy and exercise. Fortunately, your bone density doesn't get into the range of being abnormal or too low until you are much older, if it happens at all.

Fun Factoid: Keeping your bones strong for your entire life is a complicated process and depends on a lot of things. You need enough calcium and vitamin D in your diet to keep minerals in your bones. Vitamin D comes also from sun exposure, but you need healthy kidneys to turn this vitamin D into active vitamin D. You also need to exercise. Stress from exercise will strengthen bones. Women have a higher risk of osteoporosis because they are smaller, have tinier bones, and lose valuable estrogen after menopause, which shifts the balance toward thinner bones.

You have two separate parts of your skeleton that are sort of arbitrary. Early researchers in anatomy divided the skeleton into the *axial skeleton* and the *appendicular skeleton*. Your axial skeleton is a lot like the major stud walls of your house. If you knock those down, the house will be really unstable. In your skeleton, these bones are in your skull, ribcage and *vertebral column*. Think about the middle of your body and this will most likely be your axial skeleton.

Your appendicular skeleton is like the outside walls of your house. You need these bones too, but your whole house won't fall down if you knock them down. Your pelvis or pelvic girdle, the shoulder or shoulder girdle, and all the bones in your arms and legs are part of the appendicular skeleton. This figure shows the two separate parts of the skeleton:

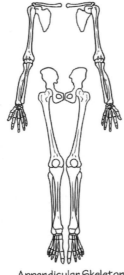

Appendicular Skeleton

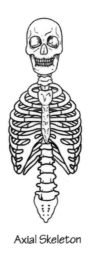

Axial Skeleton

Of course, the main functions of the skeleton are to support you and to protect your inner organs, but there's much more that the bones of your skeleton actually do. Without your bones and joints, you couldn't move in any meaningful way. You would be a blob like an amoeba or jellyfish, without a solid structure. Your bones are also where your blood cells are made inside the bone marrow, where minerals like phosphorus and calcium are stored, and where part of your endocrine system is regulated.

Bones and Their Types

You have many types of bones that are necessary for the support of your body weight and for the protection of your insides. When you think of a bone, you probably think of a long bone, like the kind a dog enjoys chewing. This type of bone would be one you'd see in your arms or legs; however, these are just a single type of bone. There are so many different bone shapes in your body. These help to shape you and have different functions based on their own unique shape.

This figure shows you what the five main bone shapes are and their names:

Types of Bones

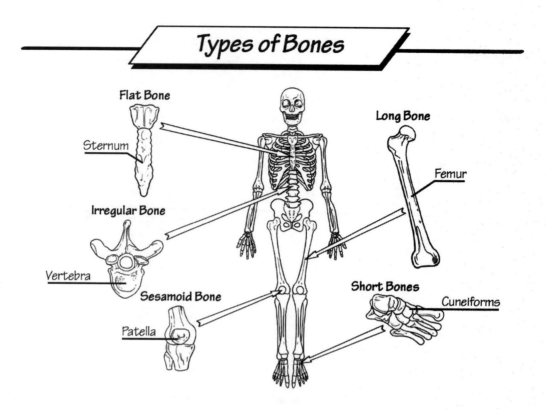

Flat bones are the main protective bones for your insides. Flat bones are seen in your skull and protect your brain. The sternum and ribs are also flat bones, even though the ribs look like they might be long bones. Ribs are actually flat when you see them up close, just like the ribs you eat for dinner. Pelvic bones are flat as well, and spread out in order to protect your pelvic structures. These act like shields to protect as much of your insides as possible.

Long bones are the ones you might think of the most as being stereotypical bones. These are the ones that help you most with regard to the movement of your arms and legs. They are long and narrow, most of them being relatively round in diameter along the shaft of the bone. The femur is the longest bone in the body and the finger bones are also considered long, even though they are actually physically short. Your appendicular skeleton is where most of these long bones are located.

Short bones are shaped like a cube. They are not just short but are also wide and stubby. Your ankle and wrist bones are short bones that help to provide joint stability. They don't do a lot of movement themselves; but in your wrist, for example, all the short bones together and the joints between them allow the wrists to be very moveable and flexible. This figure shows the short bones in your foot and ankle:

(right foot, lateral view)

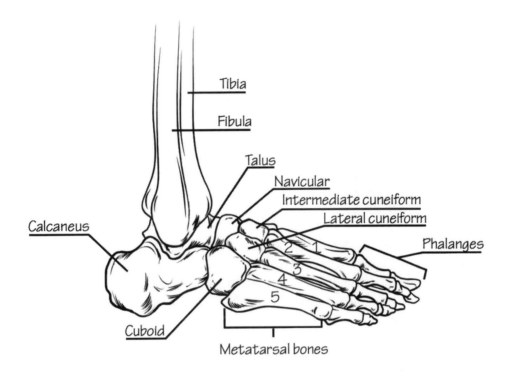

Tibia

Fibula

Talus

Navicular

Intermediate cuneiform

Lateral cuneiform

Calcaneus

Phalanges

Cuboid

Metatarsal bones

In this picture, you can see a lot of short bones, like the *calcaneus, talus, cuboid, cuneiform bones* and *navicular bone*. You can also see where the *metatarsals* and *phalanges* are long bones rather than short bones, even though they are kind of short in the span of things. This is because we call bones short if they are both short *and* wide.

Irregular bones don't have any specific shape, don't have any similar properties to each other, and are shaped according to their unique functions. The vertebrae in the spine are irregular. They have a vertebral body that is certainly short and stubby like short bones, but they also have a lot of projections coming from the vertebral body that make them irregular bones. Some pelvic bones are also irregular, like the *ilium, ischium* and *pubis*, which are fused together to make the whole pelvic structure.

Sesamoid bones are small and generally spherical. There are so many of these that they aren't always counted among the total bones you have. An added problem is that different people have different numbers of these bones, which pop up randomly inside tendons and near joints. They help to reinforce the tendons and allow for protection of these structures so they don't wear down. The *patella*, or kneecap, is a large sesamoid bone trapped within the patellar tendon. It helps improve knee mechanics by holding the tendon in one place further out from the knee joint itself, so it is more effective at extending the knee.

Bones Are Tissues

Sure, bones don't look like much. They don't seem to do anything obvious to the naked eye, so it's easy to think of them as rocks of calcium inside you that just hold you up and help you move. Bone itself, however, is a dynamic tissue structure that is always remodeling itself, depending on the stresses placed on it through your everyday activities.

An example of the dynamic nature of your bones is the *bone marrow*. Located in the middle of your bones, the marrow is a hive of activity where your blood cell components get made all the time. Minerals inside your bones that give them strength and density are added and removed from bone every day. This is also a dynamic process.

There are two types of bone you might see when looking at it with the naked eye. If you think of a long bone, for example, like the trunk of a tree, you'd see it isn't the same density when you compare the outside to the inside. The outside of a tree is very dense and has bark on it. This is necessary for the strength of the tree. If the whole tree trunk were like this, though, water and nutrients couldn't travel up into its soft interior to get to the top of the tree and its leaves.

In the same way, the outer part of the bone, or *cortical bone*, is dense. It is also called *compact bone* because of its high density. The interior of bone is called *spongy bone*, *trabecular bone*, or *cancellous bone*. The bone here is full of holes and is not dense at all.

Cortical and cancellous bone are made from the same types of cells, but they are arranged differently inside the bone. The outside connective tissue layer in bone is called the *periosteum*, which isn't made of bone at all. The periosteum is composed of elastic fibrous material, such as collagen. It also contains blood vessels and nerves. This next figure shows how these parts of bone are layered:

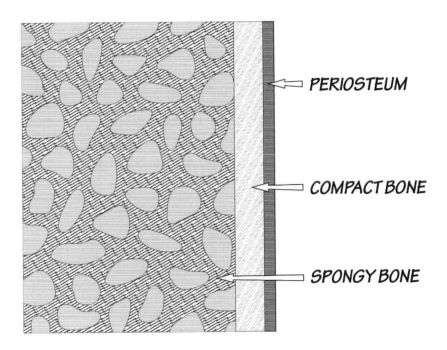

PERIOSTEUM

COMPACT BONE

SPONGY BONE

Bone is one of the tissue types we call connective tissue. In the next section, we'll talk about how bone is remodeled over time through the actions of building cells and breaking down cells of bone, called *osteoblasts* and *osteoclasts* respectively. Both of these cell types come from very early bone cells called *osteogenic cells*. These osteogenic cells are called *stem cells* in bone. They are very immature cells that haven't declared themselves to be of any particular type. As they mature or grow up, they often become osteoblasts that develop the ability to make new bone.

Bone remodeling is similar to remodeling your bathroom. You want to change something about your bathroom, so you send in workers to tear down the parts of your bathroom you don't want. In the bone, these cells are called osteoclasts. Without them, there wouldn't be room for the cells to add new bone tissue. In your bathroom, these builders who add new parts to your bathroom are the ones that install the wood, plaster, and fixtures necessary to overhaul the bathroom and make it look nice. In the bones, these worker cells are called osteoblasts.

The osteoblasts make a matrix of collagen protein and add calcium phosphate salts and other minerals to it in order to toughen the bone. Calcium phosphate salts are like the drywall or plaster on your walls. They fill in the stud wall segments and make the wall smooth and solid. The interesting thing about bone that is different from your bathroom is that when your bone gets built up, the worker cells, or osteoblasts, build all around themselves until they get trapped inside their own work. It is then that they become fully mature, stop building, and turn into osteocytes.

Most of your bone cells are these semi-dormant osteocyte cells. These cells are trapped inside spaces called *lacunae*—all surrounded by bony tissue. They aren't completely dormant because they still shore up the minerals in the bone and secrete enzymes, even though they don't divide or proliferate in any way. Osteocytes are not completely round. They have long fingers or extensions coming out of them that communicate through bony channels called *canaliculi* to other osteocyte cells nearby.

In bathroom remodeling, unless you have a spare bathroom, you can't afford to gut the whole thing and then start over. You can't have the workers tear down all the fixtures before later sending in the rebuilding workers to put the new ones back in. The process is gradual, so you can still maybe use the toilet while the remodeling is taking place. Your bones have the same issue; you still need to use them as the remodeling happens. Cells are broken down and built up at the same time so that you can still be functional during the remodeling process.

Instead, everything happens at once. There isn't a drastic change like there would be if everything were torn down and then rebuilt, but the whole bone structure stays functional just as it is changing all the time through the side-by-side activities of the osteoclasts and osteoblasts. These bone cells use the blood and nerve supply brought into these structures to provide the nutrients to continue the remodeling process.

Bone Development in Childhood

If it seems like bones are busy in your adult years, things are really hopping in the growing-up years. Babies start making bone within a few weeks of conception, a process that never really stops. The basics of the skeleton are laid down by the eighth week after conception, even though these bones aren't very calcified and are not particularly strong. These fetuses have mostly cartilage and connective tissue in their skeleton, but it doesn't take long before *ossification*, or bone mineralization, starts.

Bone development continues throughout adulthood. Even after adult stature is achieved, bone development continues whenever the bone is fractured or injured in any way.

Fun Factoid: Have you ever heard of rickets? Kids with rickets have such weak bones that they are bow-legged and have all sorts of bone deformities. The main reason for this is vitamin D deficiency, which is still a public health problem all over the world. This is why most of the milk you drink is fortified with vitamin D. This way, you get your calcium and vitamin D in one gulp. In parts of the world where people of darker skin complexions live in higher latitudes and the sun isn't as strong, they are at a greater risk of rickets because dark skin doesn't absorb as much sunlight as light-skinned people. This is because the dark pigmentation of their skin blocks the making of vitamin D from sunshine.

There are two different types of ossification, or bone mineralization: *intramembranous ossification* and *endochondral ossification*. What is the difference between them, and why are they important to bone development?

Intramembranous Ossification

Intramembranous ossification means "between membranes." It is an important way bones get strengthened during fetal development. It is also the kind of ossification, or bone strengthening, that happens if you break a bone.

There are important cells called *mesenchymal stem cells* that lie inside spaces between fractured ends of bone and in the middle of the bone. These stem cells are immature cells that turn into osteoblasts. A few of these mesenchymal cells collect together into a tiny, dense cluster called a *nidus*. The formation of a nidus means that these mesenchymal cells stop dividing and change their shape to become bigger and rounder. They build up their protein-making capacity within the cells, turning themselves into what are *osteoprogenitor cells*, or osteogenic cells. These cells develop further into osteoblasts.

Now, the real work begins. Osteoblasts get busy making the collagen fibers needed to provide the matrix needed to lay down calcium phosphate salts. This matrix is also called *osteoid*. It is a lot like widely spaced netting or mesh that just needs to be filled in by calcium phosphate salts in order to give it strength. The osteoid mesh layer gets rapidly ossified by the osteoblasts adding calcium salts wherever they can. They get a little overzealous, though, and trap themselves inside as osteocytes.

Endochondral Ossification

Endochondral ossification also happens in the fetus, and is part of fetal bone development. The difference between intramembranous ossification and endochondral ossification is that with endochondral calcification, cartilage is part of how the bone gets made. The making of bones in this way is really important in how long bones get longer and in how the fetus grows in total length and size. As bones grow in childhood, this same process is a big part of how this is accomplished.

The natural progression of endochondral ossification starts with *hyaline cartilage*, which is another type of connective tissue cell, followed by the formation of *bony cartilage*. This is actually the main way that new bone gets made in children. Hyaline cartilage is the model in which the bone structure is roughed out first before getting calcified. When building your bathroom, the workers rough out what they want by putting in stud walls, and then filling it in with plaster or sheet rock. It makes a lot of sense to do it this way.

There are worker osteoblasts that surround the diaphysis or main length of the bone in a long collar. This is where they make the dense compact bone outside the main long bone. The perichondrium is on the outer surface and helps provide the nerves and nutrients the building cells need to function. This perichondrium is the supply chain through which the necessary tools and supplies get inside the bone.

The cartilage deep within the growing bone has a different fate. It tends to degenerate over time in order to allow osteoblasts to come in and form the looser cancellous bone. There are clusters called *primary ossification centers* scattered throughout this new cancellous bone. Ossification starts at the middle of the length of bone and proceeds out to both ends. At the same time, those pesky osteocytes are always cleaning things up and breaking down unnecessary areas in order to make the *medullary cavity*.

This medullary cavity is the deepest and innermost part of the bone. This is very important because it's where blood cells are made for the entire body. This process, called *hematopoiesis*, is absolutely necessary for the production of nearly every cell your blood needs in order to function.

In children, a lot of the main ossification happens in the *epiphysis* of long bones, which are near the ends of each bone. These are like other areas of bone formation except that they're really hyperactive when it comes to making new bone and adding on the length of the bone. As the child reaches his or her full height, these epiphyseal areas completely ossify, the cartilage disappears in some cases, and no further bone lengthening can happen. A small amount of cartilage hangs on as articular or joint cartilage, and another small part stays on as a line between the epiphysis and the diaphysis. This becomes the *epiphyseal plate*, or growth plate.

This figure shows what endochondral bone ossification looks like:

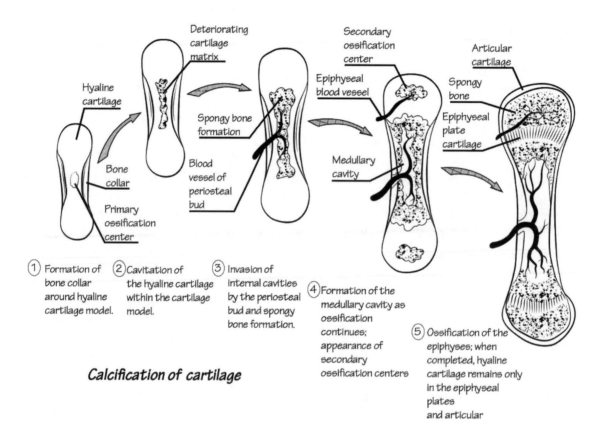

① Formation of bone collar around hyaline cartilage model.

② Cavitation of the hyaline cartilage within the cartilage model.

③ Invasion of internal cavities by the periosteal bud and spongy bone formation.

④ Formation of the medullary cavity as ossification continues; appearance of secondary ossification centers

⑤ Ossification of the epiphyses; when completed, hyaline cartilage remains only in the epiphyseal plates and articular

Calcification of cartilage

Bone Growth and Lengthening in Children

Bones do not grow all over from one end to the other. There are hot zones where almost all of the new bone is created. This figure shows what a typical long bone and its different areas look like:

Bone Anatomy

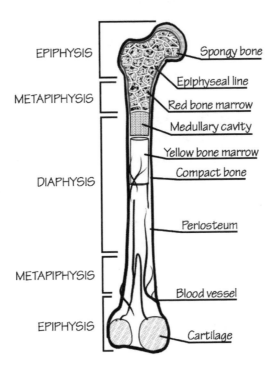

EPIPHYSIS

Spongy bone

METAPIPHYSIS

Epiphyseal line

Red bone marrow

Medullary cavity

DIAPHYSIS

Yellow bone marrow

Compact bone

Periosteum

METAPIPHYSIS

Blood vessel

EPIPHYSIS

Cartilage

The epiphyseal plate is composed of four zones. Cartilage is first made on the outer, or epiphyseal side, while this same cartilage is ossified, or strengthened with calcium salts, on the diaphyseal side, resulting in lengthening of bone in these areas. As you go from each of the four zones, you will see a progression from cartilage formation to maturation, to proliferation of early bone tissue, and finally to calcification of the cartilage on the diaphyseal side. The end result is that the bone gets longer. This process stops, of course, when you reach your adult height. Bones in this area will still be able to widen, however.

Bone Remodeling

Bone remodeling isn't something that only happens to children. It happens throughout your life. Mature bone is removed just as new bone is formed. The end result is the reshaping or replacing of damaged bony areas in order to create strong bone that is better able to withstand the pressures of daily living. Those who do not exercise or put extra pressure on their bones through weight bearing will not build or remodel bone as fast or as completely as those who do participate in stressing their bones through exercise every day.

Remodeling is really aggressive in childhood. Babies will replace their entire skeleton within the first year of life. This is a much smaller effort in adults, who turn over about ten percent of their bone every year through remodeling. If this remodeling goes awry in any way, osteoporosis can occur and the bones will be too thin. In most cases of osteoporosis, the osteoclast activity is too great and the osteoblasts can't catch up. Medications to treat osteoporosis will help reduce osteoclast activity.

A Look at the Axial Skeleton

The axial skeleton is that part of the skeleton made from the skull, ribcage and vertebral column. It consists of about 80 bones, or about just six different types of bones, including the vertebral column, the sternum, the ribcage, the *hyoid* bone, the *middle ear ossicles*, and the skull bones. Connected to the axial skeleton is the appendicular skeleton. Together, they make up the entire human skeleton.

Cranial Bones

In the axial skeleton, there are multiple flat bones that make up the skull. The skull protects the brain and is made from several infant bones that become fused into a single flat bony skull. It is a fairly strong interconnected bone. This figure shows the skull, or cranial bones, connected to each other by boundary lines:

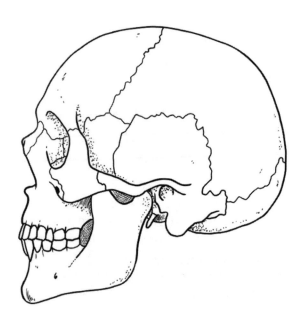

The human skull consists of the facial bones and the cranium. The cranium is the part of the skull that protects the brain, while the facial bones act to form the nose, cheeks, eyes and mouth. There are fourteen facial bones that are all connected to one another, with the exception of the mandible, which is separate from the facial bones to allow for chewing. The most important facial bones are the nasal bone, the cheek bone (also called the *zygomatic bone*), the *maxilla* (which forms the upper jaw), and the *mandible*, which connects to the maxilla.

Because the brain has not reached its maximum size at the time of birth, the skull needs to be made from several plates that eventually grow with the brain and fuse when the brain reaches its full size sometime in early childhood. The anterior and posterior *fontanelles*, or soft spots, in a baby's skull are where these bones haven't fused together completely. As brain growth tops out in size, these bones can then fuse together to make *suture lines* between them.

Rib Cage

The rib cage is made up from twelve pairs of ribs and includes the sternum. There are twenty-five separate bones in the rib cage if you include the sternum. The rib cage acts to protect the heart and lungs from damage, and has muscles attached to it that act as breathing muscles along with the diaphragm. This figure is a close-up look at the rib cage:

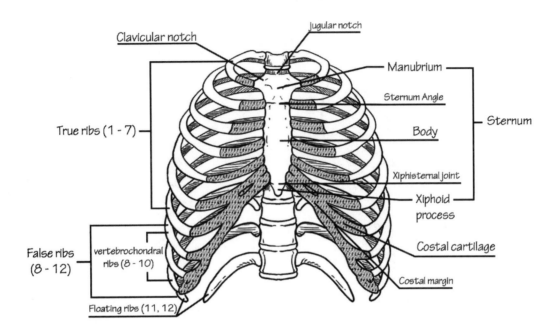

The ribs are crescent-shaped and wrap around the body, from the vertebral bodies to the sternum in the anterior part of the chest wall. The upper seven pairs of ribs attach directly to the sternum using costal cartilage. These are also known as the "true ribs." The 8th through 10th ribs have cartilage that connect them together and that stretches to connect these ribs to the sternum and upper ribs as one solid group. The 11th and 12th ribs are "floating ribs" because they are attached only at the spinal column and do not wrap all the way around the chest.

The Spinal Column

The spinal column is made of thirty-three individual but interconnected bones that are essentially stacked on top of each other. Of course, they don't just sit there; they are connected with strong ligaments and muscles that hold them together. The spinal column has many functions. It bends to some degree in all directions to allow for flexibility of the spine. The strong muscles attached to it maintain posture and allow us to lift things with our trunk.

The spinal cord itself, with all the nerve supply to the body, is highly protected by the vertebral column, or spinal column. All the spinal nerves exit out the side of the spinal column through what are called *intervertebral foramina* in order to innervate the entire body below the neck.

The spine isn't stick-straight. It has natural curves. Inward curvature toward the body is called *lordosis*, which is mainly seen in the lumbar and cervical areas. Outward curvature away from the body is called *kyphosis* and is naturally seen in the lumbar and midthoracic areas. This figure shows normal spinal curvatures:

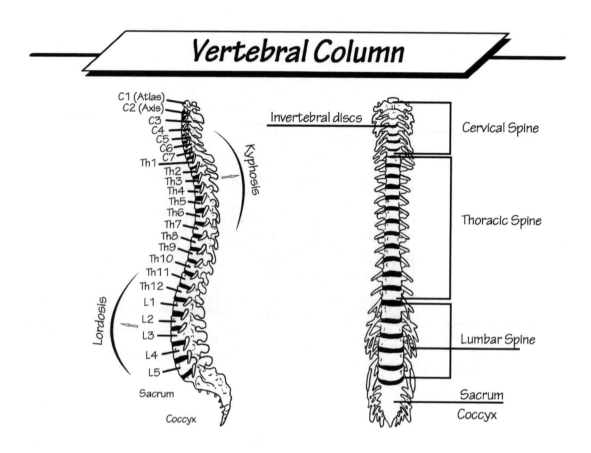

Sideways curvature is called *scoliosis*, which isn't ever normal and sometimes needs surgery or braces to correct. Outward curvature that is too great is called kyphosis, or "hunchback," while inward curvature is called lordosis, or "swayback." Muscle and posture changes will pull these vertebrae out of alignment, and vertebral body fractures will also be uneven, leading to abnormal spine curvatures.

Each of the three vertebrae are numbered from top to bottom, starting with one. There are six different regions, also from top to bottom. These regions are the cervical, thoracic, lumbar, *sacral* and *coccygeal* segments. Each has slightly different regions and functions. Let's look at these:

The cervical region helps to support, flex, bend and rotate the neck. These are small vertebrae, labeled C1 to C7. The upper two vertebrae, the *axis* and *atlas*, are very unique and have different functions than the rest of the cervical vertebrae.

The thoracic vertebrae are those that have ribs attached to them. For this reason, they're labeled T1 to T12 for every one of the twelve ribs. They don't participate much in range of motion, but have spinal nerves coming out of the lateral foramina to supply *dermatomes* (areas of skin that connect to a single nerve in the spine) on the trunk.

The lumbar vertebrae are in the lower back. These are large and thick vertebrae to best support the rest of the body above this level. There are five of these, labeled L1 to L5. They are able to rotate slightly but are more able to flex forward so we can pick up things off the floor.

The *sacrum* is below the lumbar area. It is made of five sacral vertebrae that are generally not separate, but are fused together to form a plate that communicates intimately with the rest of the pelvis. The spinal cord does not reach below about L2, but there are fibers that extend down and leave small foramina in these fused vertebral segments. These nerve fibers are called *the cauda equina*.

Fun Factoid: *Why is this lower bunch of nerve fibers called* the cauda equina? *The first part is easy. It is in the caudal, or "tail end," of you, so the term* cauda *makes sense. The second word,* equina, *refers to horses, because the entire structure looks a lot like a horse's tail.*

The *coccyx* is very small and is made of four fused vertebrae. Your coccyx is the same thing as your tailbone. Its main role is to attach muscles and ligaments to help support the floor of the pelvis. It plays no role in spinal nerve activity in the vertebral column.

The vertebral bones are irregular bones. As mentioned, they have a drum-shaped vertebral body with each body stacked on the other, separated by intervertebral discs that act as cushions between the bones. The discs act like a radial car tire. They have an outer ring, known as the *annulus*, which has bands of fibrous tissue that support the vertebrae much like the tread on a tire. The bands attach between the bodies of the vertebra. Inside the disc is a gel-filled center known as the nucleus, which acts much like the tube of a tire.

Discs act like coiled springs. The fibers of the annulus are able to bind the vertebrae together against the resistance of the elastic gel-filled nucleus. The nucleus acts like a ball bearing so that, when a person moves, it allows the vertebral bodies to roll over the gel, which is compressible. The gel-filled nucleus is made up of mostly fluid. This fluid is absorbed in the nighttime as you lie down, and it is pushed out during the day as you move around.

This next figure shows the unique processes that come out of the vertebral bodies. These processes unite with processes above and below them so that there are foramina (which means "windows" that have spinal nerves exiting through them):

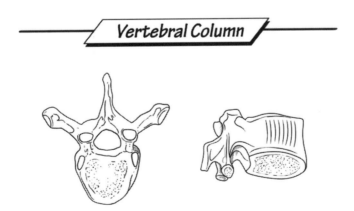

Vertebral Column

With increasing age, the discs increasingly lose the ability to reabsorb fluid and become flatter and more brittle. This is why we get shorter as we age. There are also diseases, such as osteoporosis and osteoarthritis, that can affect the vertebrae. Arthritis causes bony spurs called *osteophytes* to develop. Strain and injury can result in bulging or herniation of the discs, which is a condition where the nucleus pushes out of the annulus, compressing nerve roots and causing back pain. This next figure shows what these different disc issues look like:

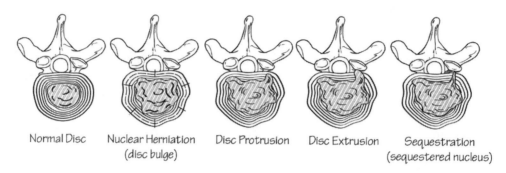

| Normal Disc | Nuclear Herniation (disc bulge) | Disc Protrusion | Disc Extrusion | Sequestration (sequestered nucleus) |

The Appendicular Skeleton

The appendicular skeleton is the part that supports your arms and legs. It also includes the pelvic girdle and the shoulder girdle, which are the different bones that connect the extremities from the axial skeleton.

Of the 206 bones in the human skeleton, 126 are a part of the appendicular skeleton. Functionally, the appendicular skeleton is involved in locomotion, or walking (lower extremities only), and in the manipulation of things (upper extremities).

The appendicular skeleton is made from mainly long bones, which accounts for the fact that these are long structures. As we have just discussed, this part of the skeleton forms during development, using cartilage as an intermediate structure, and becomes firm using a process known as endochondral ossification that makes cartilage turn into bone.

Anatomists divide up the appendicular skeleton into six segment or regions. The pectoral girdles are made from just a pair of clavicles and a pair of *scapulae*. They aren't a part of the ball and socket shoulder joint entirely; instead, they support the muscles that give the shoulder some strength. This next figure shows the pectoral, or shoulder girdle's, location:

Human Skeleton System (Pectoral Girdle) Anatomy

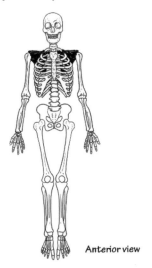

Anterior view

The arms and forearms are grouped together and are made of just three bones on each arm. The humerus is the upper arm bone, while the *radius* and *ulna* are the two lower arm bones. Next comes the hands and wrists. There are 54 bones in total, which include the eight carpal bones per wrist, the five *metacarpals*, five proximal phalanges, four intermediate phalanges, and five distal phalanges per hand. This next figure shows the arm bones:

Bones of the Upper Extremity

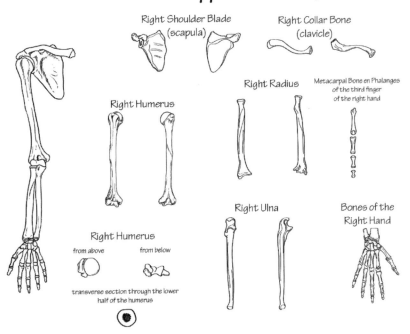

Right Shoulder Blade
(scapula)

Right Collar Bone
(clavicle)

Right Humerus

Right Radius

Metacarpal Bone en Phalanges
of the third finger
of the right hand

Right Humerus

from above   from below

transverse section through the lower
half of the humerus

Right Ulna

Bones of the
Right Hand

In the lower extremities, there is the pelvic girdle, thighs and legs, and feet and ankles. The pelvic girdle is just the fused pelvic bones that are, in adults, just called *pelvic regions*. An important structure is the *acetabulum*, which is the socket the head of the femur fits into, to make a ball-and-socket in the hip joint area.

Fun Factoid: *It's a bad thing if you break your hip at any point, but it's especially bad if you break the bone very close to the ball of the femur itself, which is called the* head *of the femur. This is because the blood supply to this area isn't good to begin with, and gets even worse after you have a break. Without nourishing blood to heal the fracture, the head of the femur just degenerates. This is part of the reason why it common to simply replace the hip joint with an artificial joint after these types of fractures.*

The thigh and legs are made from the large *femur* in the upper thigh, the right and left *patella* in front of the knee, and the right and left *tibia* and *fibula*. Other than the patella, each of these bones is a long bone that participates in weight bearing. Next comes the ankles and feet. There are 54 bones in total. There are a total of fourteen *tarsal* bones, ten metatarsals, ten proximal phalanges, eight intermediate phalanges, and ten distal phalanges. This figure gives you an idea of what the lower leg bones look like:

Bones of the Lower limb: Posterior view

Understanding the Joints

Without the different joints, bones could not easily move next to each other in order to create complex movements in the body. When you think of a joint, you might think of your knee or elbow, but these represent just a few of the many joint types that exist in your body. Some joints have free movement, while others don't move at all but still connect one or more bones to each other.

There are different classifications of joints, named according to how much movement each joint allows. There are three broad classifications of joints, known as *fibrous* joints, *cartilaginous* joints, and *synovial* joints. They are very different from one another and have their own sub-classifications. Let's look at these three main types:

- **Fibrous joints:** These joints move very little and are really more designed to cement two or more bones together. The three separate subtypes include the sutures that allow for a tiny bit of movement between two bones, such as the cranial bones of the skull. In babies, these are much more moveable than they are in older adults. *Syndesmoses* are somewhat moveable but really aren't designed for that purpose. The radio-ulnar and talo-fibular joints are syndesmoses that connect the lower arm and lower leg bones, respectively. They allow the bony ends to slide together as the extremities are moved or twisted. A *gomphosis* involves very little movement and is the joint that connects the tooth to the tooth socket. This figure shows the different fibrous joints:

Fibrous Joints

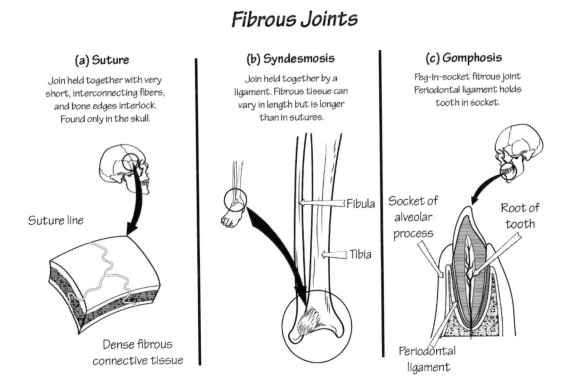

(a) Suture

Join held together with very short, interconnecting fibers, and bone edges interlock. Found only in the skull.

Suture line

Dense fibrous connective tissue

(b) Syndesmosis

Join held together by a ligament. Fibrous tissue can vary in length but is longer than in sutures.

Fibula

Tibia

(c) Gomphosis

Peg-in-socket fibrous joint Periodontal ligament holds tooth in socket.

Socket of alveolar process

Root of tooth

Periodontal ligament

- **Cartilaginous joints:** These have either hyaline cartilage or *fibrocartilage* connecting the two bones. There is more movement with these joints than there are with fibrous joints. The two types of these joints are called *synchondroses* and *symphyses*. The epiphyseal plate, or growth plates, inside long bones are examples of synchondroses, while a symphysis is what you see in the front of the pelvis between the pubic bones, where the pubic symphysis is located. This figure shows these two cartilaginous joints:

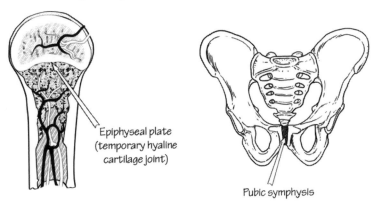

Epiphyseal plate (temporary hyaline cartilage joint)

Pubic symphysis

- **Synovial joints:** These are the workhorse joints of the body and are the ones that really allow movement between two or more bones. Without synovial joints, we wouldn't move very much at all. These are lubricated joints that have a synovial cavity. The two ends of each bone are encapsulated into a tough, fibrous articular capsule that encases the space between the two bony ends. Inside this articular capsule is joint fluid that allows for movement between the two bones with less friction. The ends of the bones are also coated with a slippery cartilage that reduces friction. The elbow and knee are both good examples of articular joints. This figure depicts what a synovial joint looks like:

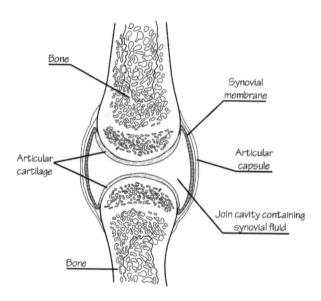

Synovial Joints Are Not All the Same

Synovial joints are similar to one another in that they all have the same basic structure, with a joint capsule, joint fluid, and cartilage on the bony ends that allow for smooth movement between the joints. But there is a difference between the elbow joint, which can bend and extend only, and the shoulder joint, which moves in all different directions. This leads to different sub-classifications of synovial joints that are separated from one another by their appearance and function. Let's look at the different sub-types of synovial joints:

- **Pivot joint:** This joint has the ability to rotate within a ring made of ligaments and the bones themselves. A good example is the joint between the axis and atlas of the cervical spine (which is the C1 and C2 vertebrae). You couldn't turn your head well if you didn't have this unique ability to rotate at the level of this joint. A similar thing is seen near the elbow at the proximal radioulnar joint that helps you turn your hand from being face up (*supination*) to face down (*pronation*).

- **Hinge joint:** This is similar to the hinges of a door. You can swing a door open or closed, but that's about it. With joints like the elbow, which is a hinge joint, you can only bend or flex it. This kind of joint has a C-shaped end at the tip of the humerus into which a piece of the ulna fits. That's what allows these two bones to extend or flex when you want them to. Your ankle, knee, and interphalangeal joints between the bones of your fingers and toes are all hinge joints.
- **Condyloid joint:** This is also called an *ellipsoid* joint. With these joints, one bone has a shallow depression on the end of it and the other one has a rounded end. This allows for not just flexion and extension of the joint but also side-to-side movements. Your knuckles, or metacarpal phalangeal joints, and the wrist joint are of this type. It's kind of like a joystick on a video game that can move up, down, right or left, but not necessarily in every direction. Every time you spread your fingers apart or make a fist, you are using these four directions.
- **Saddle joint:** This is the kind of joint that looks like it sounds. It's shaped like a saddle on one end of the bone, with a similar rounded end on the other bone that allows for a specific type of movement. You can see how this looks by moving the base of your thumb at the first carpometacarpal joint. The thumb can't move in every direction but it can do things like cross over to the opposite side of the palm when you oppose your thumb. Your *sternoclavicular* joint between sternum and clavicle doesn't move much but is also a saddle joint.
- **Plane joint:** This is also called a *gliding joint* in which the surfaces between the joints are relatively flat and similar in size. This helps each joint slide across one another in just about any direction, although most of these joints don't move very much, sometimes because there are constraining ligaments that prevent movement in every direction. Examples of this joint are the intercarpal and intertarsal joints in the wrist and ankle. Rotate your wrist to see how these bones can move in several directions.
- **Ball-and-socket joint:** This is the joint with the greatest degree of movement. One bone has a cavity called a *socket*, while the other has a ball at the end of it. The ball fits into the socket and ligaments hold the whole thing together. Like a regular ball and socket, these joints move fluidly in many directions. The shoulder joint and the hip joint are both ball-and-socket joints.

This figure shows roughly what each of these synovial joints looks like:

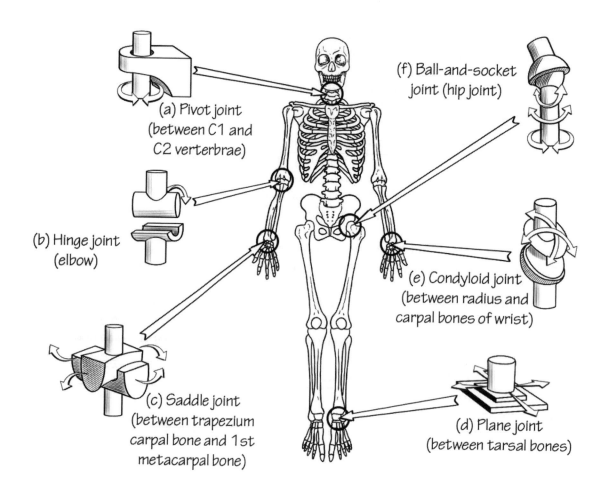

(a) Pivot joint (between C1 and C2 verterbrae)

(b) Hinge joint (elbow)

(c) Saddle joint (between trapezium carpal bone and 1st metacarpal bone)

(d) Plane joint (between tarsal bones)

(e) Condyloid joint (between radius and carpal bones of wrist)

(f) Ball-and-socket joint (hip joint)

The Role of Cartilage

Without cartilage covering synovial joint surfaces, the two rough bones would rub against each other when in motion. The movement certainly wouldn't be fluid or smooth, and it might even be painful. Enter the beauty of cartilage, which acts to provide a smooth, lubricated surface between bones.

Cartilage is an extremely resilient elastic tissue. It provides a rubber-like padding that protects and covers the joints and long bones in the body. It also participates in the formation of surfaces related to the rib cage, the bronchial tubes, the ear, the nose and the intervertebral discs. It isn't as hard or rigid as bony tissue, but it is much less flexible and stiffer than muscle.

Because of its rigidity, cartilage often serves the purpose of holding tubes open in the body. An example of this is the rings of the trachea, such as the *carina*, the *cricoid* cartilage, and the *torus tubarius*, which is found at the opening of the *Eustachian tube*, the outside of the ear, and the openings or holes of the nostrils.

Cartilage comes from specialized cells known as *chondrocytes*. These cells have the important job of making a large amount of extracellular matrix that is made from collagen, elastin fibers, and abundant ground substance that is high in *proteoglycan*. All of these are proteins that are tough but flexible. By themselves, these cartilage components aren't slippery; but when exposed to the fluid inside the joint, they tend to become more slippery because they make for a smooth surface.

There are three types of cartilage: fibrocartilage, hyaline cartilage, and elastic cartilage. Each differs in the relative amounts of proteoglycan, elastin and collagen inside. The types with the most elastic elastin fibers tend to be more flexible than fibrocartilage, which doesn't have much stretch to it.

Cartilage is considered *avascular*, meaning that it doesn't contain a lot of blood vessels. This means it can't be too thick or it won't get the nutrition it needs to be healthy tissue. The nutrition that cartilage layers get comes from the diffusion of nutrients that the chondrocytes embedded in the deeper layers supply to the entire cartilaginous surface. The downside is that because there isn't a lot of blood flow to cartilage, when it gets damaged it doesn't repair itself well, and takes a long time (if ever) to heal after an injury.

To Sum Things Up

Your bones are like the stud walls of your house. They aren't easily seen but are very important in holding your body up and in protecting all your internal parts. Rather than being static structures like the two-by-fours in your house, though, bone is dynamic. It is being remodeled all the time and is a good repository for calcium, magnesium and phosphorus in the body. There are different shapes of bones that have various functions. Connecting your bones are your joints, which allow bones to move, bend and twist near one another.

The missing part of all of this is that these structures can't move by themselves. Take a look at any real skeleton and you can see it just hangs there, incapable of any movement, even though the joints are there to allow for movement. This is where your muscles, ligaments and tendons come in. You need these structures to hold the bones in place, and to attach those bones to all the muscles of your body. Your muscles provide the real energy you need to get moving. In the next chapter we'll talk about the muscular system, and how it works with the skeletal system to make you the mover and shaker you are most of the time.

Chapter Seven: Your Muscles and How They Work

Locomotion Happens!

Muscles are all about locomotion, or movement. Bones are nice but kind of useless without the muscles to move them. Fortunately, you have a muscular system that allow you to move in all directions, walk, and even maintain your posture. The bones do most of the protecting of your internal organs, but the muscles of your back and abdomen also do this. They give your bones some shape, providing a real example of putting meat on your bones.

As you will see in this chapter, muscles do not act by themselves. You need your nervous system to provide the right signals to the muscles in order to allow for movement. You also need to have some way of attaching muscle to bone. This is where tendons come in. They are dense connective tissue structures that essentially provide the means to cement a muscle on one end to a bone and, further down the line, cement the same muscle to another bone. When the muscle contracts, the joints often bend, and the body shows some kind of movement.

Of course, this is a simplified way of looking at things. Some muscles don't move anything, but just remain in their tensed state so you can maintain your posture standing in a cafeteria line, for example, without collapsing on the floor while waiting. You might not actually move any part of your body, but your muscles are still tense and keeping you upright the whole time. Similarly, there are tiny muscles in your face that don't attach to any bone, but are still tense as you smile or frown at the person next to you in line.

Your muscles work best through the simple mechanism of having good body mechanics. When you lift something with your forearms, the biceps muscle in your forearm (your "Popeye muscle") contracts to do this. It would be useless, however, if there wasn't some way of relaxing the muscles on the other side of the upper arm so they aren't opposing your biceps muscle.

In this section, we will look at skeletal muscles from a microscopic standpoint, so that you can see deep inside a muscle fiber and better understand how it does all its contracting. We'll also talk about tendons and their purpose, as well as how muscle actions work to move the body in different ways. Finally, we'll talk about different muscles and how they act with one another in order to support and move your body.

A Look Inside Your Muscle Cells

Skeletal muscle is the muscle you think of when it comes to moving your body. However, this is just one of three different muscle types you have. The other types of muscle include *smooth muscle* and *cardiac muscle*. Skeletal muscle is called *striated muscle* and is under voluntary control. What this means is that your motor nervous system sends fibers from your brain to tell the muscles to move or not. Smooth muscle and cardiac muscle are not voluntary muscles, so they contract without you thinking about it. They are similar to skeletal muscle in many ways, but have their own unique functions.

A single skeletal muscle is made up of many bundles of cells known as muscle fibers, or *fascicles*. The muscles and different bundles of muscle fibers are encased in connective tissue layers known as *fasciae*. Muscle fibers are in the shape of a cylinder and their cells contain more than one nucleus. Muscle fibers are, in turn, made from *myofibrils* inside the muscle cells themselves.

The myofibrils are made from *myosin* and *actin* filaments, which are repeated in units called *sarcomeres*. Sarcomeres are the basic units of the muscle fiber, and are the whole reason a muscle fiber contracts. As you'll see, sarcomeres look unique under a microscope. There are lines you can see called *striations*. This is why both cardiac muscle cells and skeletal muscle cells are called *striated muscles*.

Let's look first at how skeletal muscles are arranged. Around each muscle fiber is a connective tissue sheath called the *endomysium*. There are up to 150 muscle fibers in one fascicle. Around a single fascicle is a sheet of connective tissue called the *perimysium*. The *epimysium* is a larger piece of connective tissue that surrounds the muscle itself. This figure shows how a muscle is organized:

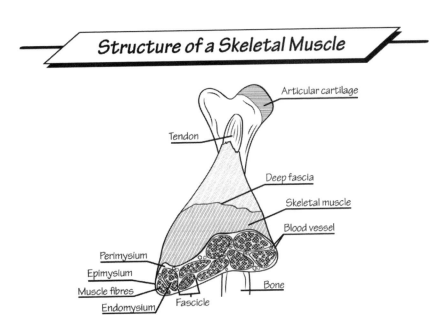

It takes more than a long stretch of myofibrils to have a muscle cell contract. The cell membrane is called the *sarcolemma*. It is important in allowing ions to travel into and out of the cell, which starts the contraction process. There is also a different kind of endoplasmic reticulum in each muscle cell called the sarcoplasmic reticulum. Its job is to be a storehouse for calcium ions. You'll see in a minute why this is so important to the muscle cell. This figure shows you what the myofibril and a single sarcomere look like:

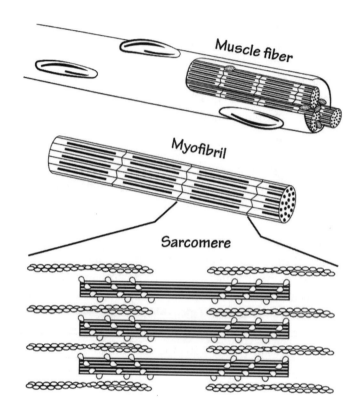

So, what's in a sarcomere that makes it able to lengthen or shorten when muscles relax and contract? Sarcomeres are called *repeating units*. They're like the cars of a train that line up in a long line to make the entire train. At the end of each sarcomere is a thick protein disc called a *Z line*. This is what glues each sarcomere to the one next to it, much like there are structures that connect train cars together. It's the Z line you see under the microscope as the lines we call *striations* in striated muscle.

Sarcomeres have the amazing ability to lengthen or shorten. Actin is attached at either end of the sarcomere to the Z lines, but the myosin sections are not fixed to anything in a permanent way. They don't just sit there but instead stay connected to the actin protein fibers in a dynamic way.

Where the myosin and actin fibers overlap in a cross-section of a sarcomere, there is a thickened line called an *A line* that you can also see under the microscope. The *H zone* is the part on a cross-section of a sarcomere where there are only myosin filaments present. This figure shows a close-up look at a sarcomere when it is relaxed and when it contracts:

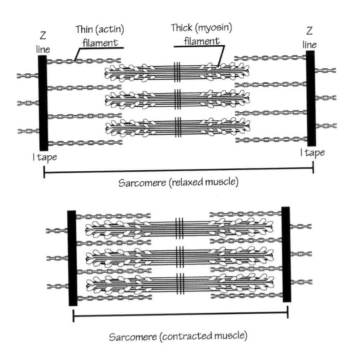

Sarcomere (relaxed muscle)

Sarcomere (contracted muscle)

How a Muscle Cell Contracts

As you can imagine, there are several steps along the way from you deciding to move a muscle and the muscle itself obeying your mental command. Let's break it down into the different steps on a tiny microscopic level. You'll have to use your imagination to see how the contraction of one sarcomere doesn't do much, but when all of the sarcomeres from all of the muscle fibers in a single muscle do the same thing, you can really get some work done.

The first step involves translating your mental or neurological message into each muscle fiber that needs to contract. We'll talk about the nervous system later but, with regard to muscles, each muscle fiber is connected to a motor nerve that has traveled to it from your central nervous system. One *motor unit* is the same as one nerve fiber, plus all the different muscle fibers connected to it. Your nervous system is a lot like a 'robocall' telephone system, where one source calls a bunch of people at once, giving them the same identical message. This is what a motor unit looks like:

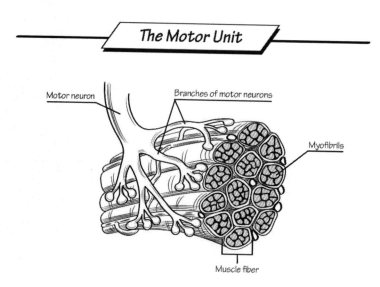

The point where the nerve cell interacts to send the signal to the muscle cell is called the *neuromuscular junction*. While the robocaller just has to dial your number to send you their message, the nerve cell does this through chemical messaging. The tip of the nerve has a chemical messenger called *acetylcholine* that is sent into a small space between the nerve and the muscle, called the *neuromuscular junction synapse*. This chemical attaches to receptors on the muscle fiber and the message is sent. This is what the neuromuscular junction looks like:

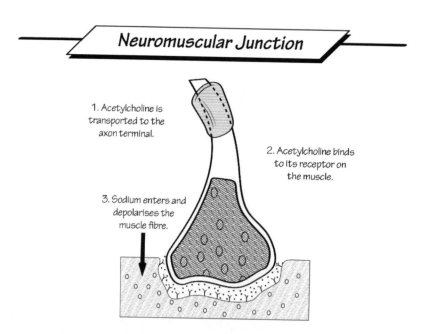

Fun Factoid: *Nerve gas is really bad for you, mostly because of the effect it has on the neuromuscular junction. Nerve gas blocks the enzyme that breaks down the acetylcholine at the neuromuscular junction. This means that this messenger continually turns on the muscle contraction, so they go into spasm and you can't move, breathe, or control your bowels or bladder.*

At the level of the muscle fiber, ion channels in the sarcolemma are opened up to allow a lot of sodium into the muscle cell. This is called *depolarizing* the cell membrane, an electrochemical change in the cell. Remember that sodium ions are charged, so when one of these ions rushes into the cell, it changes the charge across the membrane. In the muscle cell, there are deep channels called *T tubules* that send this electrochemically charged signal to the interior of the cell where the sarcoplasmic reticulum is located.

Once the sarcoplasmic reticulum gets the signal, it dumps all of its calcium ions into the cytoplasm of the muscle fiber. This is where the myofibrils are located, which is a good thing because this is exactly what these myofibrils need in order to be activated to contract the muscle fiber.

Actin and myosin, the two main proteins that make up the myofibril, have a strange interaction with each other. There are binding sites on the actin chain that want to bind to the myosin chain, but this is prevented by two proteins that are collectively called *the blocking complex*. The proteins are called *troponin* and *tropomyosin*. It turns out that, if calcium is available, it binds to this complex, changing it in some way to uncover the binding sites necessary for actin and myosin to connect to one another. The two chains are now free to form a cross-bridge between them. Here's what these chains and the cross-bridges look like:

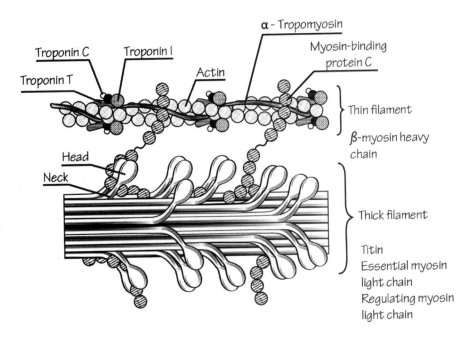

Of course, if the cross-bridges just stayed connected, nothing would happen to the muscle. It needs another step to allow the bridges to break and reform in ways to shorten the muscle. This is where ATP comes in. It binds to myosin, breaks the cross-bridge, and causes the tiny heads on the myosin to swivel so that they attach to the next nearest binding site. The way these heads are oriented means that the reconnection to the next site actually drags the actin along the myosin chain, shortening the entire sarcomere a little bit.

All of this activity shortens the actin and myosin segments of each sarcomere in a progressive fashion, causing the Z lines at the end of each sarcomere to get closer together. With thousands of sarcomeres along the line doing the same thing, the muscle fiber shortens, and therefore your entire muscle shortens. When the electrochemical signal from the nerve is no longer active and the robocaller hangs up, there are pumps that pull all the calcium back into the endoplasmic reticulum, disconnect actin and myosin from one another, and lengthen the muscle. Like after the annoying robocaller stops ringing your phone, your muscle forgets the whole thing until the next time *another* signaling message is sent to the muscle fiber.

Muscles Contract in Different Ways

When you think about muscle contraction, you probably think about lifting something with your arms, watching the muscles contract and shorten, effectively bending your elbow so you can lift the object. This certainly is a type of muscle contraction, but it isn't the only type of muscle contraction you can have. It's called a *concentric muscle contraction* because it shortens the muscle.

In *isometric* muscle contractions, the muscle tenses but doesn't lengthen or shorten. When you make a muscle with your biceps but don't move the elbow at all, this is an isometric contraction. The two types of isometric contractions are *yielding* contractions and *overcoming* contractions. A yielding contraction involves holding something that is resisting you, like a heavy weight. An overcoming contraction is when you push against something immovable like a wall.

In *isotonic* muscle contractions, the muscle length will change in some way. These are divided into *eccentric* and *concentric* contractions. An eccentric contraction is one where the muscle lengthens and the angle made by the joint increases. It isn't the same thing as muscle relaxation. The muscle instead lengthens in a controlled way. A concentric contraction is when the muscle shortens and the angle made by the joint decreases. This figure shows what the different muscle contractions look like:

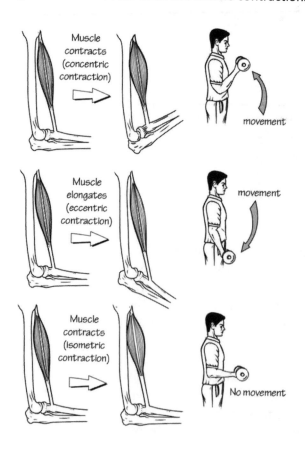

How Muscles Attach to Bones

Muscles generally don't attach directly to bones, but use tendons that interweave between the ends of the muscle in order to attach to the periosteum of a bone. There is connective tissue around the muscle (the epimysium) that continues on as a stiff collagen-containing tendon to affix to the proper bone on both ends of the muscle. Tendons can be short or long. Some have fluid-filled sacs called *bursae* near the rough spots, where even a strong tendon would have too much wear and tear over time and with repeated use. Bursae help to lubricate the tendon over certain areas, usually near joints. This figure shows what a bursa looks like in the shoulder:

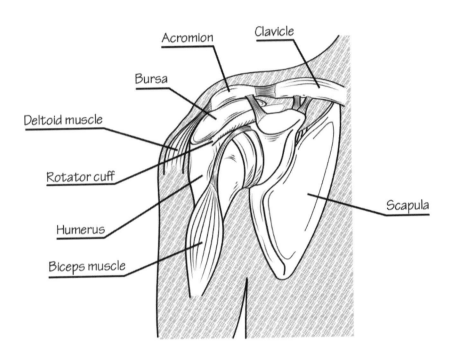

Tendons must be strong enough to withstand tension placed upon it by the action of the muscle. Tendons and ligaments are similar; they are both made from collagen. The biggest difference between the two are that tendons attach muscle to bone, while ligaments connect two bones together without any muscle involved.

When looked at under the microscope, tendons look like long bands of collagen protein packed together in linear bundles. While more than 85 percent of a tendon is collagen, the rest is made from other proteins and substances, such as elastin.

Fun Factoid: Most super-stretchy people who can do things like bend over backwards easily and dislocate their joints do this for a reason. Many of these folks have a connective tissue disease called Ehlers-Danlos Syndrome. There are several types of this genetic disease but most involve a defect in the ability to make normal collagen. Their elastic fibers are okay, so they are very stretchy. This is actually dangerous because it can affect the collagen in the internal organs, leading to heart and blood vessel diseases, along with a chance of an early death.

A single collagen molecule is about 300 nanometers in length and about 1-2 nanometers in width. The *fibrils*, or bundles of collagen molecules, are about 50-500 nanometers in total diameter, which means there are up to 500 collagen molecules per bundle. In tendons, the bundles are about 10 millimeters in length but are interwoven to make a tendon that is as long as necessary. If you look at a tendon up close, these are the structures you'll see:

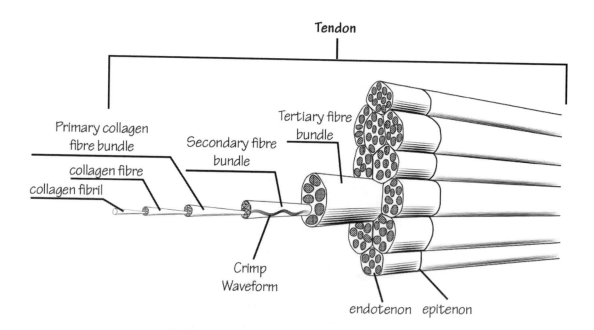

Not all muscles attach to true tendons, nor do all of them attach to bones. Some muscles instead just attach to other muscles. The connective tissues segment between two connected muscles is called an *aponeurosis.* Aponeuroses are made from the same collagen as tendons, but are flat sheets of collagen that allow two nearby muscles to be attached to one another.

Origin and Insertion

In a little bit, we'll talk about a lot of different muscles. Before we do this, though, there are a few terms you should understand. Two of these terms are called the *origin* and *insertion* of a muscle. When you think of the average muscle, you might notice that there are generally two ends (although some muscles will split off to involve more than one muscle bundle). The origin of the muscle is defined as where the muscle attaches at its most proximal end, or the part closest to the center of the body. The insertion of the same muscle is where the muscle attaches at its most distal end, or the part furthest from the center of the body.

Flexion and Extension

When you study the anatomy of the muscles, it is important to know the ways the body moves. The trunk and extremities can flex and extend. *Flexion* is defined as the bending of the body or a joint, such as the wrist or elbow. *Extension* is the opposite of flexion. It involves the straightening of the joint so that it creates a 180-degree angle. The hip can extend to become completely straight through the process of extension.

We've already talked about hinge joints. The basic action of hinge joints is to flex and extend a joint. They really can't do anything but that. True flexion and extension involves no rotation of the joint whatsoever. Just think "bend and flex" and "straighten and extend" and you've got these terms figured out.

Abduction, Adduction, and Circumduction of Joints

A couple of other terms you should know are *abduction* and *adduction*. Adduction involves the movement of an arm or leg joint toward the center of your body. For example, when the arm is brought to the side, it is called adduction. When the arm is brought away from the body, it is called abduction. The same can happen in the lower extremities, as the leg can abduct and adduct through movement of the hips. *Circumduction* only happens with ball-and-socket joints such as the hip and shoulder. It involves the circular movement of the limb in nearly 360 degrees around the joint.

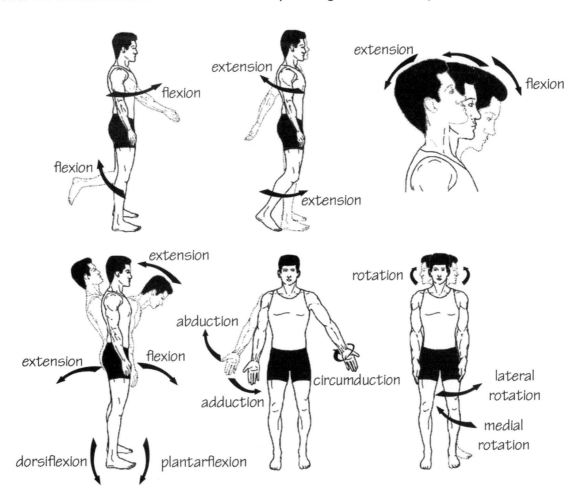

Medial and Lateral Rotation

Medial and lateral motion are other ways that an extremity joint can move. When a shoulder undergoes *medial rotation*, the shoulder is rolled forward toward the upper chest. When it is rolled away from the body, it is known as *lateral rotation*. The hip can also rotate medially and laterally. These are the main joints that are capable of medial and lateral rotation. Hinge joints can't really do these kinds of movements.

Protraction and Retraction of Facial Muscles

Protraction involves the forward projection of a facial structure. It is seen whenever the mandible of the jaw moves forward and anterior to the orbital plane. It is a term commonly used in dentistry and is seen when you stick your jaw out. It means protrusion or extension of the facial structures.

Retraction involves the act of drawing back or being drawn back. When you draw your mandible or chin inward toward the center of your head, you're retracting it.

Pronation and Supination of the Arm or Leg Joints

Pronation and *supination* are related to the movement of the foot or hand. When you turn your palm upward, this is called *supinating* your wrist. When you turn your palm downward, it is called *pronating* your wrist.

Pronation and supination also occur with respect to the body itself. You can lie prone on a table, which is also referred to as pronation. It means that you are face down with respect to the table or bed. Supination of the body involves the condition of being face up or supine.

Eversion and Inversion of Your Foot

Eversion and *inversion* are terms that are mainly describe your ankles and feet. For example, the foot can invert when the sole of the foot is turned inward. It everts whenever the sole of the foot is turned outward. An inversion injury happens when the ankle is forcefully twisted inward as a person steps into a crevice, off a curb, or into a hole.

The Muscles

In this section, we will talk about a lot of different muscles. While there are more than 650 named muscles in the human body, you should not attempt to memorize these unless you plan to become an anatomist yourself. Instead, pay attention to the main movers of a joint. Many joints have a main mover muscle that does most of the work, and a lot of secondary muscles that help do the job.

Another thing to pay attention to are clusters of muscles that work together. There are, for example, your forearm *flexor* muscles that all bend the elbow. Some muscle groups are in certain compartments, such as the anterior or posterior compartment of the lower leg. Try to see which muscles are found in which compartment, and imagine what the group of muscles do as a whole rather than the specifics of each muscle by itself.

If you really want to study muscles because you are planning to be an orthopedic surgeon, for example, you will need to know very specific things, like which exact tubercle on a given bone a muscle inserts into, which nerve supplies that muscle, and where the blood supply comes from for each muscle. These are important things to know if you are going to do surgery on a muscle, but not as important for everyday understanding of the muscles themselves. For this reason, it doesn't make a lot of sense to clutter your mind with the specific tiny nerve branch or artery that supplies any of the muscles we will talk about. Instead, we will talk about what the muscles are, what they look like, and what they do.

The Muscles of Facial Expression

Your muscles of facial expression are those that make you smile or frown, depending on the situation. Few of these muscles are attached to the bones of the face. Instead, they reside in the *subcutaneous* tissue beneath the skin, many of them attaching to the skin itself or to each other with connective tissue between them. They contract in a myriad of ways in order to make all of the familiar facial expressions you know. This is a picture of your face with all of these muscles separated out:

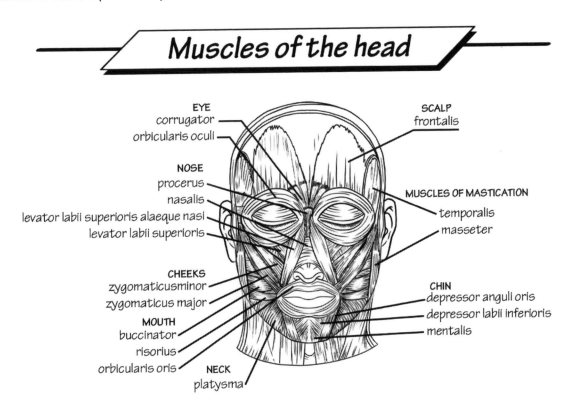

There are three main categories of facial muscles under this broad category. These are the orbital, nasal, and oral muscle groups. Let's look at the orbital muscles first:

- **Orbicularis oculi:** This is the muscle around the eye and in the eyelid itself. There are three parts that gently or tightly close the eyelids and help drain tears. It extends like a circle around your eye.
- **Corrugator supercilii:** This is a tiny muscle hidden behind the orbicularis oculi muscle. It is the muscle that creates wrinkles on the bridge of your nose as you draw your eyebrows together. It is known as the *frowning muscle*.

The nasal group is the cluster of facial muscles that help you wiggle your nose. There are three muscles that help with this function:

- **Nasalis:** This is the muscle on the side of your nose. It has a part along the *nares*, or nose holes, that helps you flare and compress them.
- **Procerus:** This is the topmost muscle of the nose, lying just beneath the skin. It originates in the nasal bone and inserts into the lower part of the forehead. It is the muscle that pulls your eyebrows downward so that you can cause transverse or horizontal wrinkling above your nose.
- **Depressor septi nasi:** This is a muscle below the nose that attaches to the *nasal septum* and down to the maxilla just above your front teeth. If you can do this, try to pull your nose toward your mouth. This is the muscle you are using. The idea is that it opens your nares.

The oral group of facial muscles is the most important muscle group in this area among humans. You need these muscles in order to smile and frown. There are two major muscles that accomplish this, plus a few tiny ones that help.

- **Orbicularis oris:** This is the muscle that opens and closes your mouth. It surrounds the mouth after attaching to the maxilla and other cheek muscles. It also has attachments in the skin and lips. When you purse your lips, you are using this muscle.
- **Buccinator:** This is a deeper muscle found between your maxilla (cheekbone) and your mandible (jawbone). It is the muscle that forms your cheeks, and keeps your cheek from bulging out when you eat or from having food collect in your cheek pockets.

The Muscles of Mastication

Mastication is basically chewing, and you can't do that without the right muscles for the job. There are four separate muscles that do this. Here's what they look like:

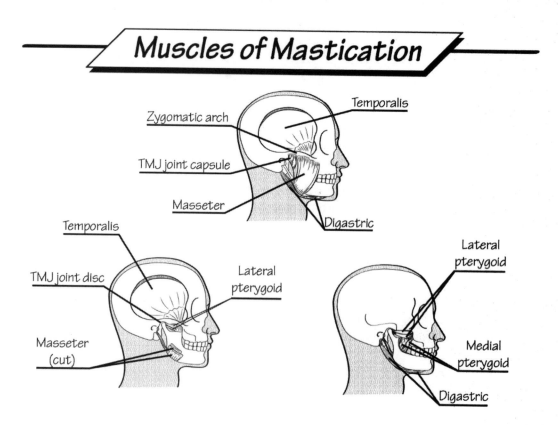

The four main muscles of mastication are:

- **Masseter:** This is a major muscle for chewing, and the main one you use when you eat. There are deep and superficial parts to this muscle, which also covers the *temporalis* and *pterygoid* muscles. It originates in the zygomatic bone and the zygomatic arch, and inserts into the mandible so you can chomp down on something.
- **Temporalis:** This is a muscle on the side of your skull, reaching back to above your ear. While it doesn't sound like it would be related to chewing, it is. It originates on the *temporal* bone on the skull but inserts onto a *lateral mandibular process*, so that it both closes the jaw and retracts the mandible.
- **Medial pterygoid:** This is a quadrangle-shaped muscle with deep and superficial heads. It helps to close the mouth but isn't the main muscle that does this.
- **Lateral pterygoid:** This is a triangle-shaped muscle that also has two heads, a superior one and an inferior one. It projects the jaw forward and is the main muscle that does this. You need both of your lateral pterygoid muscles to jut your chin forward. It also allows side-to-side movement of the jaw during chewing.

The Muscles of Your Neck

The muscles of your neck have a lot of things to do. Not only are they responsible for keeping your neck upright, but they must also allow you to move your neck in all directions. Additionally, some of these help do things like swallowing and speaking. There are a lot of neck muscles, which are further divided into different regions.

The first region involves the *suboccipital* muscles, which are in the back of your neck. There are four of these muscles. They are called suboccipital muscles because they lie beneath the *occiput*, or back of the head.

- **Rectus capitus posterior major:** This is a large muscle on the lateral posterior part of the neck. It starts at the C2 vertebra and inserts into the inferior part of the occipital bone. Because of its location, you can see where it would extend and rotate the head.
- **Rectus capitus posterior minor:** This is a medial muscle that actually connects to the *dura major* that surrounds your brain, but has origins in the C1 vertebra and inserts into the occipital bone. Because of its connections to your internal brain structures, if it gets inflamed it might be the cause of headaches related to the neck, called *cervicogenic headaches*.
- **Obliquus capitis inferior:** This is a low-lying suboccipital muscle that is not actually attached to the cranium. It starts at the C2 vertebra at its *spinous process* and ends at the *transverse process* at C1. It helps to extend and rotate the head.
- **Obliquus capitis superior:** This is a lateral muscle in the back of the neck that starts in the transverse process of C1 and attaches at the base of the occipital bone. It is mainly involved with neck extension.

This is what these deep muscles look like. Show your understanding by coloring these four muscles within the illustration below.

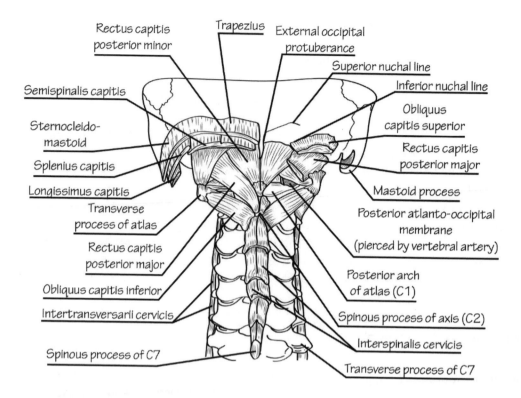

Rectus capitis posterior minor
Trapezius
External occipital protuberance
Superior nuchal line
Inferior nuchal line
Obliquus capitis superior
Rectus capitis posterior major
Mastoid process
Posterior atlanto-occipital membrane (pierced by vertebral artery)
Posterior arch of atlas (C1)
Spinous process of axis (C2)
Interspinalis cervicis
Transverse process of C7
Semispinalis capitis
Sternocleido-mastoid
Splenius capitis
Longissimus capitis
Transverse process of atlas
Rectus capitis posterior major
Obliquus capitis inferior
Intertransversarii cervicis
Spinous process of C7

Moving around to the front of the neck, you'll see the *suprahyoid* muscles located above the small hyoid bone in the front of the neck. The hyoid bone is interesting because it isn't attached to any other bone, but is attached to muscles. The goal of these muscles is to raise the hyoid bone when you swallow. These are the muscles that help do this:

- **Stylohyoid:** This is a thin muscle above the *digastric* muscle that starts at a tiny process called the *styloid process* located in the temporal bone, and attaches to the far lateral part of the hyoid bone. By pulling up the hyoid bone, it initiates swallowing.
- **Digastric muscle:** This muscle has two connected bellies. It starts at the mandible anteriorly and at the mastoid process of the temporal bone posteriorly. These two bellies attach to the hyoid bone and depresses the mandible while elevating the hyoid bone at the same time.
- **Mylohyoid:** This is a broad muscle that helps to form the oral cavity floor, so it is a major supporter of that area. It begins at the mandible and attaches to the hyoid bone. It helps to elevate both the floor of the mouth and the hyoid bone.
- **Geniohyoid:** This is a midline muscle lying beneath the mylohyoid bone. It starts just below the chin at the mandible and ends at the hyoid bone. It also elevates the hyoid bone and depresses the mandible.

Fun Factoid: The hyoid bone is tiny and isn't attached to anything but muscles. If you are strangled, though, you can end up breaking this bone. Unfortunately, most of the time a bone like this is discovered to be broken only during an autopsy after a strangulation death.

This is where these muscles can be found in the upper anterior neck:

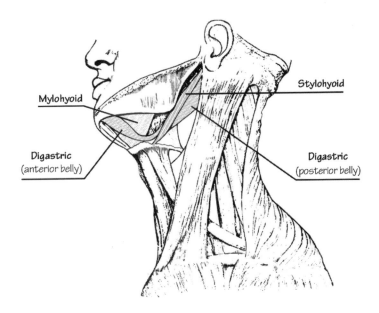

Below the hyoid bone are the *infrahyoid* muscles. There are two superficial muscles and two deep muscles:

- **Omohyoid muscle:** This is a muscle with two bellies that are connected by a tendon sheath. The inferior belly starts at the scapula and runs beneath the large *sternocleidomastoid* muscle. It attaches to the superior belly, which ascends upward to the hyoid bone. Its function is to depress the hyoid bone.
- **Sternohyoid:** This is a superficial muscle that starts on the sternum at the sternoclavicular joint. It travels upward to insert into the hyoid bone so that it can depress it.
- **Sternothyroid:** This is wider and deeper than the sternohyoid muscle. It is a deep muscle that starts on the sternum and attaches to the thyroid cartilage, where it depresses it.
- **Thyrohyoid:** This is a short and deep muscle that continues from the sternothyroid muscle. It starts at the thyroid cartilage in the *larynx* and ascends to the hyoid bone, where it serves the function of depressing the hyoid bone. It can also elevate the larynx.

This figure shows what these muscles look like in the anterior part, or front, of the neck:

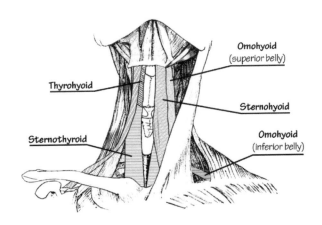

The *scalenes*, or scalene muscles, are along the side of the neck, where they help make the floor of the posterior triangle of the neck area. Their main job is to flex the neck forward. There are three of these muscles on either side of the neck. This figure shows what they look like:

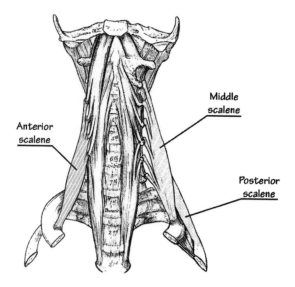

- **Anterior scalene muscle:** This is on the lateral neck beneath the sternocleidomastoid muscle. It starts along C3 to C6 at the transverse processes and attaches to the scalene tubercle, which is on the first rib. It helps to elevate this rib making in part of *inspiration*. It also laterally flexes the neck on one side but forward flexes the neck when both sides are activated.
- **Middle scalene muscle**: This is the largest of the scalene muscles. It starts along the transverse processes of C2 to C7 and also attaches to the first rib. It serves to elevate the first rib but will also laterally flex the neck to one side.
- **Posterior scalene muscle:** This muscle attaches to the transverse processes of C5 to C7 and inserts onto the second rib. It will elevate the second rib and help flex the neck to one side.

There are a couple of other neck muscles you should know about that are considered superficial. These are the *platysma* and the sternocleidomastoid muscles.

- **Platysma:** The platysma is a thin muscle on the front of the neck, that starts on the lower part of the face and extends to the upper thoracic area, where it connects to the skin of this area. It covers much of the upper chest muscles. This muscle allows depression of the mandible and wrinkling of the lower facial skin, which is not a very important function in humans.
- **Sternocleidomastoid muscle:** This is located along the side of the neck. It starts at the mastoid process behind the ear on the temporal bone and travels down to the sternum and clavicle in the upper chest. It is responsible for laterally flexing the neck, and for elevating the head if both of the muscles on either side of the neck work together.

This image shows these muscles on the front and side of the neck. Color the SCM and the Platysma muscle to show your understanding.

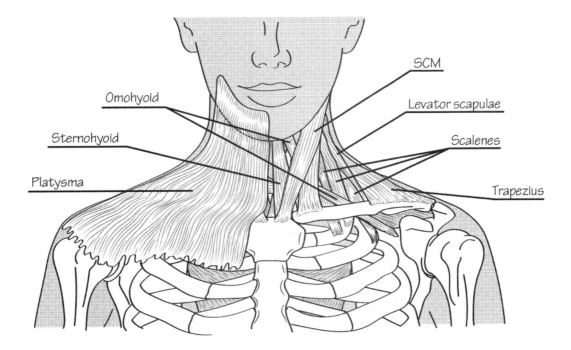

Muscles of the Back

The muscles of the back are divided into three separate groups based on how deep they are. There is a superficial, intermediate and deep group of muscles. Each of these does a slightly different thing, as you will see.

The superficial muscles of the back are just under the skin and often attach to the shoulder. Their major function is to move the shoulder in some way rather than to actually move the back. These are the major muscles:

- **Trapezius:** This is a large muscle that forms a trapezoid shape when combined with its pair on the opposite side of the upper back. It starts at the base of the skull as well as the *ligamentum nuchae* and the spinous processes of C7 through T12. It has multiple functions, including elevating the scapula during arm abduction, retracting the scapula, or pulling the scapula downward, depending on which fibers are used.
- **Latissimus dorsi:** This covers a wide expanse of the lower and mid-back. It has origins all along the spinous processes from T6 to T12, the lower three ribs, the *thoracolumbar fascia*, and the *iliac crest*. They travel along laterally and upward so that it can insert into the upper humerus of the arm. It has a major role in extending, adducting, and medially rotating the arm.
- **Levator scapulae:** This is a strap muscle that starts in the neck and travels down to the scapula. It helps to elevate the scapula, although there are other muscles that do this as well.
- **Rhomboids:** There are two rhomboid muscles, called the *rhomboid major* and *rhomboid minor* muscles. The rhomboid minor muscle sits above the rhomboid major muscle. The rhomboid minor starts on the spinous processes of C7 to T1 and attaches to the medial part of the scapula.

The rhomboid major muscle starts on the spinous processes of T2 to T5 and also attaches to the medial part of the scapula. Both of these retract the scapulae and bring them together. They also rotate the scapulae as they contract.

Fun Factoid: *Because of their location and function, the rhomboid muscles are the ones most easily strained by rowers, or members of crew teams. These are the muscles used to pull oars back, so they get overused or strained if the rower is too aggressive in training, or starts rowing without warming up first.*

This image shows you what these superficial muscles look like:

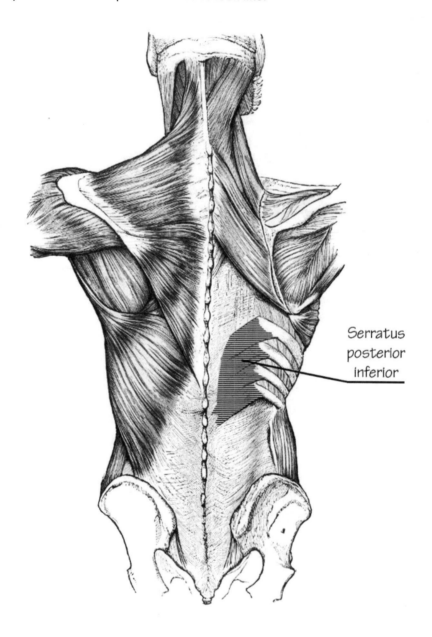

Serratus posterior inferior

The intermediate back muscles involve two major muscles. Their major role is to elevate or depress the rib cage, which means that they may partially help with respirations. This image is what they look like in the mid-back region:

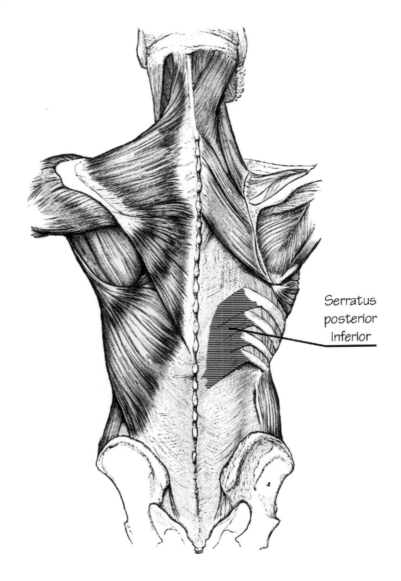

Serratus posterior inferior

- **Serratus posterior superior:** This lies deep to the rhomboids and starts along C7 to T3, extending downward and laterally to attach to the second to fifth ribs. Its job is to elevate these ribs.
- **Serratus posterior inferior:** This lies deep in the latissimus dorsi and starts at T11 to L3, where it passes laterally and upward to attach to the ninth through twelfth ribs in order to depress them.

The deep muscles of the back are also called *intrinsic back muscles*. These are the main posture muscles, so they often run along the entire back. There are three separate layers of these back muscles, which are also called superficial, intermediate and deep layers (not to be confused with the other superficial, intermediate and deep muscles of the back we've talked about. These are all are deep muscles in general, despite how they're labeled).

- **Splenius capitis:** This is a muscle on the back and side of the neck. It starts along the spinous processes of C7 to T4 and on the lower part of the ligamentum nuchae. It attaches to the mastoid process of the temporal bone and the occipital bone. Its job is to rotate the head to the same side as the muscle.
- **Splenius cervicis:** This muscle starts along the spinous processes of T3 to T6, with ascending fibers that go up to the transverse processes of C1 to C4. It also rotates the head to the same side as the muscle.

These two muscles together help to rotate and extend the head and neck area. This is what they look like:

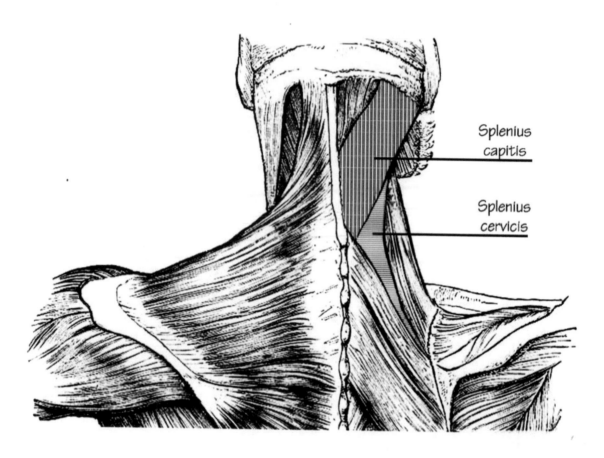

Splenius capitis

Splenius cervicis

The *erector spinae* muscles involve three intermediate intrinsic back muscles that together form a column running up and down the back. They have different attachments on one end but the same attachment on the other. This is what these muscles look like:

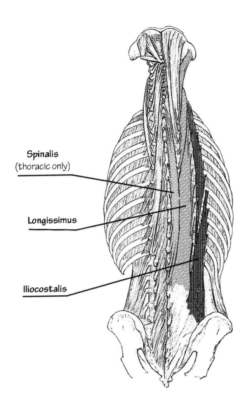

- **Iliocostalis:** This is the most lateral of the three muscles. On the lower end, it starts with a common tendon above the pelvis and ascends to insert along the posterior ribs and the transverse processes of the cervical spine. It helps to flex the vertebral column laterally and to extend the head when they are used on both sides of the body.
- **Longissimus:** This also travels on either side of the spine and lies between the iliocostalis and the spinalis muscles. It starts at the common tendon in the lower back and extends up to the transverse processes of C2 to T12, the mastoid process of the skull, and the lower ribs. It also flexes the back laterally or extends the spine and head when used together on both sides of the back.
- **Spinalis:** This is the most medial and smallest of these muscles. It starts along the common tendon and travels up to the spinous processes of C2 as well as the occipital bone and T1 to T8.

Below these muscles is the deepest area of muscles in the back. The muscles of this layer are short and often link the transverse and spinous processes of the vertebral column. This image shows what they collectively look like:

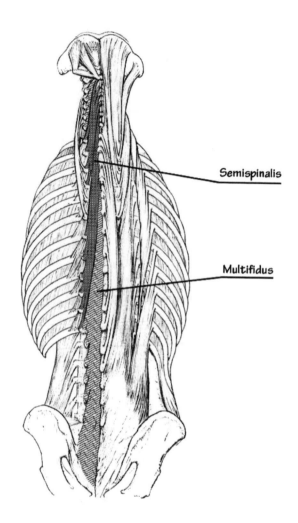

- **Semispinalis:** This muscle starts along the C4 to T10 vertebrae at the transverse processes and travel up to the levels of the spinous processes of C2 to T4, going up just four to six segments of vertebrae at a time. Some fibers go up to the skull itself. It helps to extend and rotate the head and spinal column.
- **Multifidus:** This sits beneath the semispinalis and starts in the posterior iliac spine, the sacrum, and the common tendon of the erector spinae muscles. It also has attachments all along the transverse processes on T1 to T3 and onto the articular processes of C4 to C7. The fibers go up only two to four segments of the spine before attaching to the spinous processes of the various vertebrae. Its goal is to stabilize the spine.
- **Rotatores:** These are relatively prominent thoracic muscles that start along the vertebral transverse processes, before ascending to attach to the spinous processes and lamina of the vertebra just above it. It also stabilizes the spine by having proprioceptive activity, so it helps you know if your spine is straight or not just by how these muscles tense and relax over time.

There are also tiny muscles, including the *interspinales* muscle between adjacent spinous processes, the *intertransversari* muscles that travel from one transverse process to another, and the *levator costarum*, which helps to elevate each rib.

Pectoral Muscles

Your pectoral muscles, or "pecs," are on the anterior part of the upper chest. They are considered muscles of the upper extremity because they mainly act on the shoulder joint. There are four of these muscles that often work together:

- **Pectoralis major:** This is a superficial muscle that has two heads. There is a clavicular head attaching to the clavicle, and a sternocostal head attaching to the sternum and some ribs. It attaches laterally to the upper humerus. It helps to adduct and medially rotate the arm, drawing the scapula downward and anteriorly. It also helps to flex the arm. This is what it looks like:

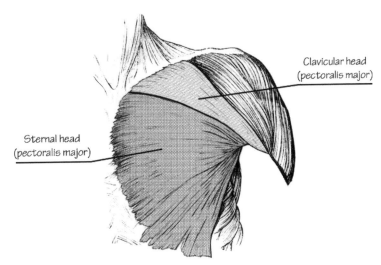

- **Pectoralis minor:** This lies beneath the pectoralis major and helps to form the shape of the anterior chest wall and the *axilla*. It starts at the third through fifth ribs and ends at the *coracoid process* on the scapula. It helps to stabilize the scapula by keeping it affixed to the chest wall.
- **Serratus anterior:** This is along the lateral chest wall and helps to shape the axilla, or underarm area. It starts along the lateral parts of ribs one through eight and attaches to the scapula along its medial border and along the part of the scapula that faces the ribs. It helps to rotate the scapula when you raise your arm above your head, and keeps this bone affixed to the rib cage.
- **Subclavius:** This is a tiny muscle beneath the clavicle, where it runs along its length and helps to protect some of the delicate nerves and vessels in the area. It also helps to anchor the clavicle and depress it.

This is what these muscles look like:

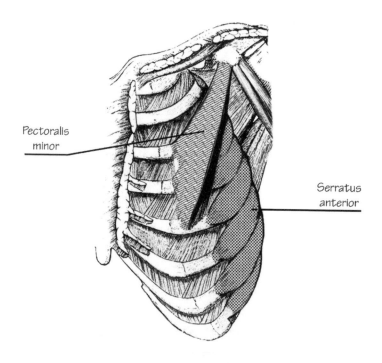

Shoulder Muscles

The shoulder muscles are also linked to moving the arm. These form the shape you see in the shoulder. There are both extrinsic and intrinsic muscles of this area. The extrinsic muscles are those we've talked about, such as the trapezius and latissimus dorsi muscles. Let's look now at some of the intrinsic muscle groups of the shoulder:

- **Deltoid muscle:** This is the muscle that most shapes the shoulder. It starts along the lateral part of the clavicle along with parts of the scapula. It attaches to the upper and lateral part of the humerus so that it both laterally and medially rotates the shoulder, as well as flexes and extends the shoulder. The middle fibers will abduct the upper arm.
- **Teres minor:** This forms part of the space through which the major arteries and nerves pass in the shoulder region. It starts along the lower scapula and attaches to the upper humerus in order to help adduct and extend the shoulder. It helps to medially rotate the upper arm as well.
- **Rotator cuff muscles:** These are a group of muscles that help to provide stability to the shoulder joint itself. There are four of these muscles that work together to do this: the *supraspinatus*, *infraspinatus*, *subscapularis*, and teres minor muscles.

This is what these shoulder muscles all look like around the shoulder:

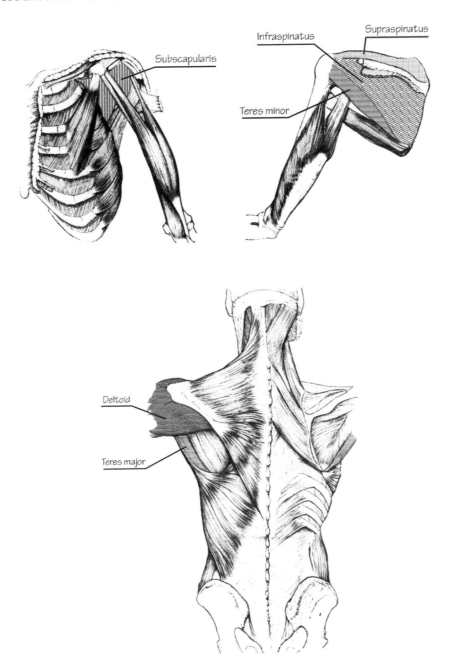

Upper Arm Muscles

These are the muscles between the shoulder and the elbow. They are divided into two compartments, the anterior and the posterior. While they are located in the upper arm, they actually do most of their work on the lower arm by helping to change the flexion and extension of the elbow joint.

There are three muscles in the anterior compartment of the upper arm. These are the *biceps brachii*, the *coracobrachialis*, and the *brachialis* muscles. The same nerve and artery supply these muscles together.

The biceps does not actually attach to the humerus, but instead attaches to the scapula. It inserts distally into the fascia of the forearm and into the upper radius, where it supinates and flexes the forearm.

The coracobrachialis and brachialis muscles lie beneath the biceps brachii. These also help to flex the forearm at the elbow, while the coracobrachialis muscle slightly adducts the forearm. Neither of these will work well by themselves, but instead help the biceps brachii carry out the action of forearm flexion. This is what these muscles look like:

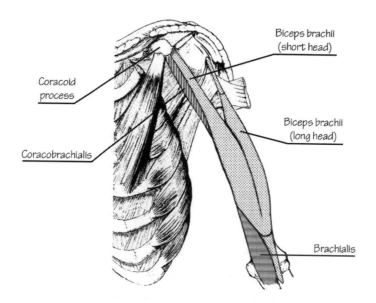

There is just one muscle in the posterior upper arm that naturally opposes the anterior muscles. Its job is to extend the forearm, or to straighten it. It starts at the upper humerus but has three separate heads that do this. There is one distal point for this muscle, which is at the *olecranon process* of the proximal ulna. This is what it looks like:

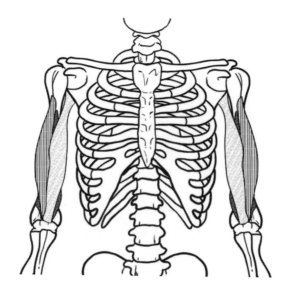

Forearm Muscles

The forearm is complex and has many different muscles in various compartments. Added to the complexity is the fact that there are often several layers of muscles.

The anterior compartment has the primary function of flexion of the forearm at the elbow joint and adduction of the arm at the shoulder joint. There are superficial, intermediate and deep layers of muscles. Most of these muscles will do one of two things: either flex the wrist or the fingers themselves, or pronate the forearm, which means flipping the forearm from the palm-down to the palm-up state.

The superficial muscles of this anterior compartment are the *palmaris longus*, *flexor carp ulnaris*, *pronator teres*, and *flexor carpi radialis* muscles. All of them have a common tendon as their origin, which is located on the medial aspect of the distal humerus. This is what these muscles look like in the forearm of the left hand:

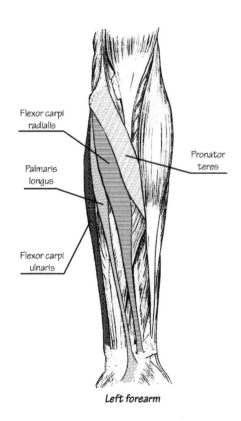

Left forearm

There is just one muscle in the intermediate compartment of the anterior forearm, which is called the *flexor digitorum superficialis*. It has two heads that start at the medial *epicondyle* of the humerus and the radius. Distally, it splits into four separate tendons that travel underneath a connective tissue band in the forearm through the carpal tunnel, where it attaches to the middle phalanx of each of the four fingers (but not the thumb). It helps to flex these fingers at the metacarpophalangeal joints and at the proximal interphalangeal joints. It also helps to flex the wrist.

There are three deep muscles in the anterior compartment of the forearm, which are the *flexor pollicis longus*, the *flexor digitorum profundus*, and *pronator quadratus* muscles. The flexor digitorum profundus helps to flex just your fingertips separately (which is really hard to do). The flexor pollicis longus helps to flex the thumb. The pronator quadratus sits between the radius and ulna and helps to pronate the forearm. This is what the deep muscles of the anterior forearm look like:

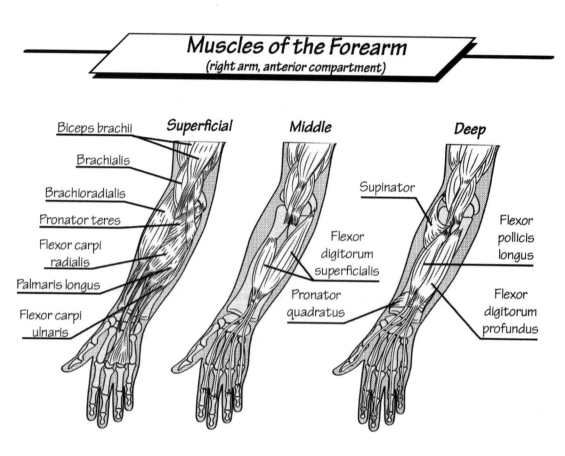

The posterior compartment has the *extensor* muscles of the hand, fingers or wrist. It has deep and superficial layers. There are a lot of muscles in this compartment of the forearm. In the superficial layer alone, there are seven different muscles. Four muscles have the same origin at the lateral epicondyle of the radius. These are the *extensor digitorum*, the *extensor carpi radialis brevis,* the *extensor digiti minimi*, and the *extensor carpi ulnaris*.

The brachioradialis is one that looks like an extensor muscle but actually is a flexor muscle. It starts at the lateral humerus distally and attaches to the distal radius. When it contracts, it flexes the elbow. The extensor carpi radialis longus and brevis muscles work together on the posterior forearm. (The term *longus* means "long" and the term *brevis* means "short.") These both abduct the wrist and help to extend it.

The *extensor digitorum communis* is the main muscle that extends the fingers by attaching distally to the medial part of the four fingers. The pinkie finger is special because it gets its own extensor muscle, called the *extensor digiti minimi*. The *anconeus* sits by itself because, while it also attaches to the lateral epicondyle, it travels just a short distance to help stabilize and extend the elbow, while also abducting the ulna when the forearm is pronated. This image shows what these muscles look like:

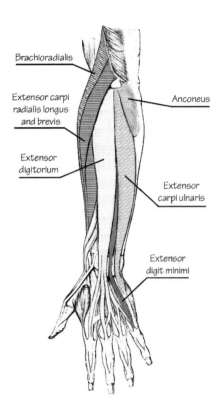

Believe it or not, there are five more deep muscles in the posterior forearm. These mostly act on the thumb and pointer finger in order to help extend them. The *supinator* does something quite different, though. This image shows these deep muscles:

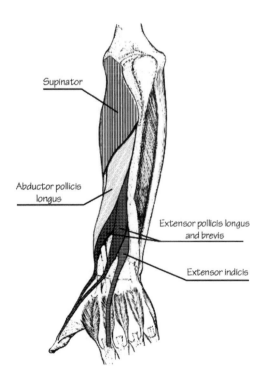

The supinator has two heads at its origin, which are the lateral epicondyle and the posterior ulna. They come together in order to insert into the posterior radius. Its job is to supinate the forearm. The other muscles of this layer are the *abductor pollicis longus*, the *extensor pollicis brevis*, the *extensor pollicis longus*, and the *extensor indicis proprius*. They act on the thumb or the forefinger. Whenever you see *pollicis*, it means the thumb is involved. If you see *indicis*, it means your index finger is involved. If you see *digiti minimi*, it's your pinkie finger.

Hand Muscles

Because the hand is so ornate and needs to perform minute movements, there are a lot of small muscles in it. You already know the extrinsic muscles, which are the forearm muscles we just talked about. This leaves mainly the intrinsic muscles that help to do all of the fine hand muscle movements. These are also separated into different groups, including the *thenar* and *hypothenar* muscles of the thumb and ulnar side of the hand, respectively.

The three thenar muscles are short and found at the base of the thumb. These form the bulge you feel in this area. The three muscles are the *opponens pollicis*, *abductor pollicis brevis*, and the *flexor pollicis brevis*. They allow the thumb to do all its different movements, including opposition, which is being able to touch your thumb to your pinkie finger. This is what they look like:

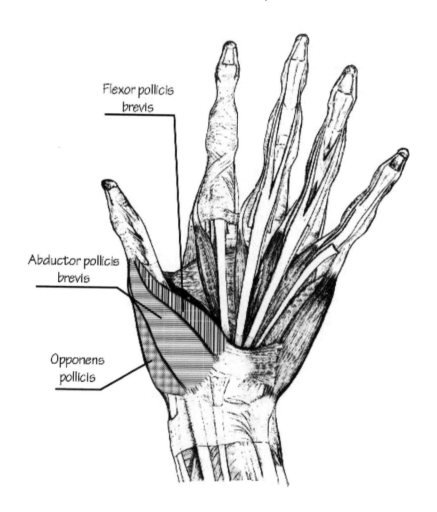

The hypothenar muscles are those that form the *hypothenar eminence*, which is the bulge on the opposite side of the hand as the thumb. Located on the palm, there is the deep *opponens digiti minimi* muscle that helps the pinkie finger reach over to touch the thumb. The others are the *abductor digiti minimi* and the *flexor digiti minimi brevis*, which also act on the pinkie finger. They look like this:

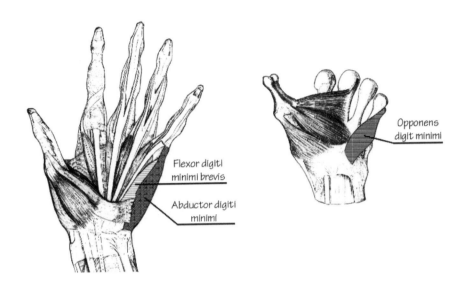

Each finger also has its own *lumbrical* muscle, which helps in finger muscle movements. They work to link the activity of the extensor tendons of the fingers to the flexor tendons. They allow for flexion of the metacarpophalangeal joints, and extend the interphalangeal joints of each finger. The lumbricals look like this:

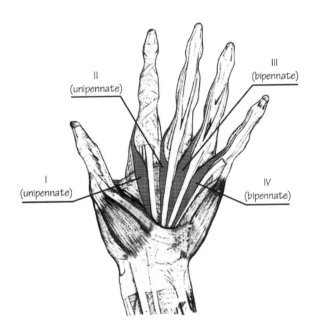

There are also dorsal and palmar *interossei* found in the palm and the back of the hand. The four dorsal interossei abduct or spread the fingers apart. The palmar interossei are buried in the palm of the hand.

There are three of these that will adduct or bring the fingers together against resistance. Together, they look like this:

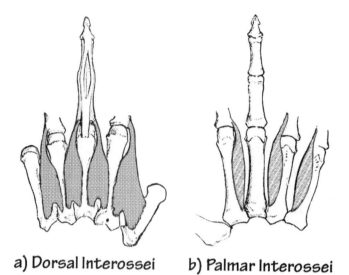

a) Dorsal Interossei   b) Palmar Interossei

The two odd muscles of the palm of the hand include the *palmaris brevis* and the *adductor pollicis*. The palmaris brevis deepens the curvature of your hand when you grip something, while the adductor pollicis is a large muscle that helps to adduct the thumb. This is what these muscles look like:

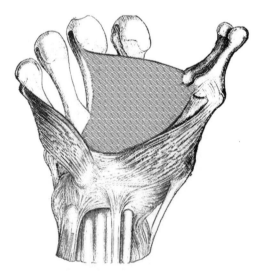

a) Adductor pollicis

The Abdominal Wall Muscles

Your abdominal muscles are more than just your abs, although those are important too. There are anterior or posterior aspects to the different abdominal muscles. The abdominal wall protects the viscera and help these organs to stay in their proper place. When you breathe out forcefully, you use these muscles to push up on the abdominal viscera inside. They are also important in the acts of defecating, vomiting, and even coughing.

There are several layers of tissue in the anterolateral aspect of the abdominal wall. The outside layer is the skin. Deeper to that are the superficial fascia, the muscular layer, and the *parietal peritoneum*.

The superficial fascia is just below the skin. There are two separate parts: the part above the umbilicus (belly button) and the part below the umbilicus. Above the umbilicus is a single sheet of connective tissue, while below the umbilicus there are actually two separate layers, called the *Camper's fascia* (which is fatty in most people) and the *Scarpa's layer*, which is deeper and membranous.

Below the superficial fascia are the abdominal wall muscles, divided into flat muscles and vertical muscles. The three flat muscles lie on the lateral side of the abdomen, one on top of another. They have fibers that run all over and in different directions in order to create a relatively impenetrable and strong layer. Each of these muscles forms an aponeurosis, or connective tissue tendon, that covers the most superficial vertical muscle, the *rectus abdominis*. As these come together in the midline, the *linea alba* is made, which extends from the *xyphoid process* at the bottom of the sternum down to the pubic symphysis.

The flat muscles include the external oblique muscle, which is quite large and has fibers running inward and downward. It starts in the ribs and extends down to the pubic tubercle and the iliac crest, where it can rotate the torso opposite to the side of the muscle. The internal oblique muscle has fibers that run inward and upward, starting at the *inguinal ligament*, the *lumbodorsal fascia*, and the iliac crest. These fibers go on to insert into the ribs, where they rotate the torso on the side of the muscle or help you bend over at the waist. This figure shows what they look like:

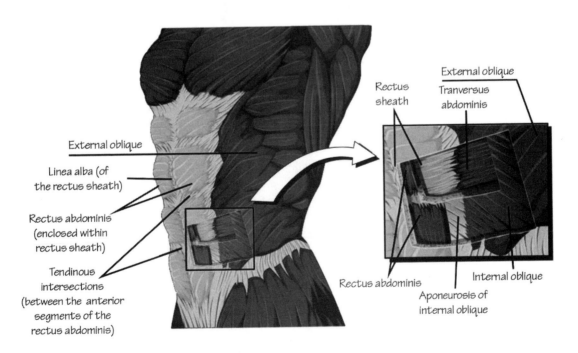

Deeper is the *transverse abdominis* muscle, which has fibers running across the abdomen from side to side. There is a fascia layer beneath it called the *transversalis fascia*. Its job is to compress the contents of the abdomen.

The two vertical muscles are the rectus abdominis and the *pyramidalis*. The rectus abdominis is the vertical muscle on either side of the midline. It is split by the linea alba and creates the *linea semilunaris* along its lateral border. There are tendons that run across the muscle in segments, giving the six-pack look you see when the muscle has been exercised a lot.

The pyramidalis muscle is small and triangular, found just above the rectus abdominis muscle. It's located just above the pubic bone and is attached into the linea alba. Its only job is to tense the linea alba in the front of the lower abdomen.

It's harder to define what the posterior abdominal wall looks like, because you can't actually see it. Even so, there are five muscles that make up this part of the abdominal wall. They are the *psoas major, psoas minor, iliacus, diaphragm,* and *quadratus lumborum*. This is what they look like:

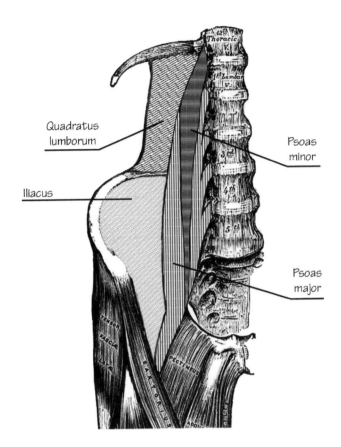

- **Quadratus lumborum:** This is the muscle along the lateral aspect of the posterior abdomen. It sits above the psoas major muscle and attaches to the iliac crest and the iliolumbar ligament, with muscle fibers running inward and upward to insert into the transverse processes of the lumbar vertebrae. Its job is to extend or laterally flex the lumbar spine. It also keeps the twelfth rib fixed during inspiration, to help the diaphragm be more effective.
- **Psoas major:** This is an important muscle near the midline that starts at T12 to L5 and travels down beneath the inguinal ligament to insert into the femur. Because of this location, it both laterally flexes the back and helps to flex the thigh.
- **Psoas minor:** This is not an important muscle; only 60 percent of people have one at all. It sits in front of the psoas major muscle and helps to flex the vertebrae, extending from the T12 to L2 level and finally attaching to the superior *ramus* of the pubic bone.
- **Iliacus:** This muscle is shaped like a fan. It's located in the lower part of the abdominal wall and combines fibers with the psoas major to make the *iliopsoas* muscle. Both of these muscles are the main ones that flex the thigh.

The Fascia Lata of the Thigh

The *fascia lata* of the thigh isn't muscular but is still important in how the thigh works. It is a deep fascia that starts at the level of the iliac crest in the pelvis near the inguinal, or groin, area. It travels downward to insert into the tibia at the bony prominences. It helps to divide the thigh into anterior, medial and lateral compartments. A major opening into this fascia layer is the *saphenous ring* lying just below the inguinal ligament. This is where the major lymph vessels and veins drain out of the leg and back into the body.

This is what the fascia lata looks like:

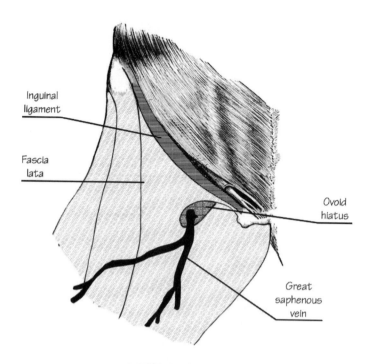

The *iliotibial tract* or *iliotibial band* (athletes call it the *IT band*) is a long and thick part of the fascia lata, extending along the lateral aspect of the thigh from the iliac tubercle to the lateral condyle of the upper tibia. The job of the IT band is to help extend, abduct and laterally rotate the hip, compartmentalize the thigh muscles, and act as a sheath around the *tensor fascia lata* muscle.

The tensor fascia lata muscle not only tenses the fascia lata but also flexes, abducts and internally rotates the hip. It starts at the iliac crest and moves down to insert along the upper lateral aspect of the thigh. It helps to brace your knee when you lift up the opposite foot. Lastly, by tensing itself, it helps to compress the deep leg veins, to allow good venous return from the lower legs. This is what the muscle looks like:

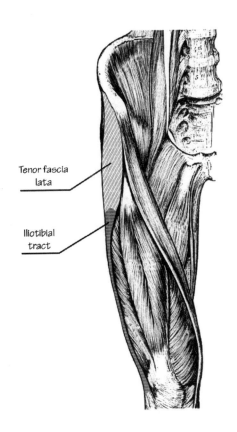

Tenor fascia
lata

Iliotibial
tract

Gluteal Muscles

Your gluteal muscles are your "butt muscles" that tend to do a lot when it comes to moving your hips. There are two groups of these muscles: superficial extenders and abductors of the hips, and deep lateral rotator muscles.

The three superficial muscles are the ones that shape your thigh and do most of the thigh extension. These are the *gluteus maximus*, *gluteus medius*, and *gluteus minimus* muscles. The gluteus maximus is the largest and strongest muscle for this purpose, but is only used when you run or climb, when more force is required.

The gluteus medius sits between the gluteus maximus and gluteus minimus. Its shape and function of abducting and medially rotating the thigh is similar to that of the gluteus minimus muscle. The gluteus minimus is the deepest of these muscles. It prevents pelvic drop of the opposite leg by keeping the pelvis secure. This figure shows what these muscles look like:

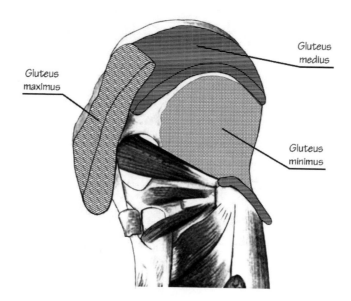

The deep gluteal muscles lie beneath the gluteus minimus. These all tend to laterally rotate the leg at the thigh level and help to stabilize the hip joint itself. This figure shows what these muscles all look like together:

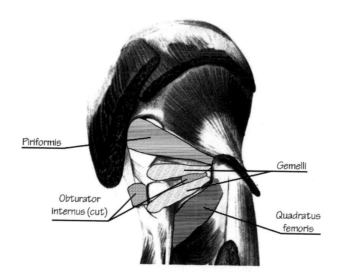

Thigh Muscles

The thigh muscles are located all around the thigh. The anterior thigh muscles are just three muscles that help to extend the knee. These are the *sartorius*, *quadriceps femoris*, and *pectineus* muscles. The distal end of the iliopsoas is located here as well and acts to flex the hip joint. Let's look at some of the thigh muscles:

- **Quadriceps femoris:** This is actually four separate muscles that act together. They are the *vastus lateralis*, *vastus intermedius* and *vastus medialis*, plus the *rectus femoris*. Together, these form a powerful muscle group used to extend the knee. They also form the quadriceps tendon where the patella attaches in front of the knee.

130

- **Rectus femoris:** The rectus femoris starts at the anterior inferior iliac spine and areas across the ilium. It then attaches to the patella along with the vastus muscles. The main difference between this muscle and the vastus muscles is that it flexes the thigh and extends the knee as well.
- **Sartorius:** This is the longest muscle you have. It starts at the level of the anterior superior iliac spine and attaches to the upper medial part of the tibia. It helps to flex the knee and both abduct and laterally rotate the hip joint.
- **Pectineus:** This is a flat muscle sitting at the base of the femoral triangle. It attaches to the pelvis and inserts along the back of the femur proximally. Its main job is to adduct and flex the hip joint.

This is what these anterior thigh muscles look like:

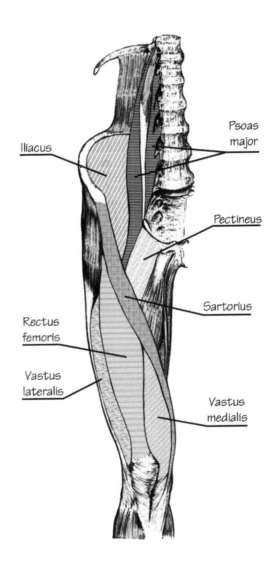

There are also medial muscles located in the medial compartment of the thigh. They collectively act to adduct the hip. There are five separate muscles in this medial group, which are the *obturator externus, gracilis, adductor longus, adductor brevis* and *adductor magnus*. This is what these muscles look like:

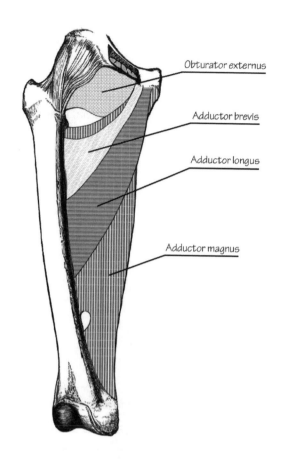

Obturator externus

Adductor brevis

Adductor longus

Adductor magnus

The posterior compartment of the thigh is the group of muscles known as the *hamstrings*. There are three hamstring muscles: the *semitendinosus*, *biceps femoris* and *semimembranosus*. In the back of the knee, they form a tendon that can strongly extend the hip and flex the knee. These are the muscles that are weak when you get sciatica, because they all get innervated by the large *sciatic* nerve.

Fun Factoid: *The meaning behind the word* hamstrings *depends on who you ask. Some say it relates to an Old English word, not* ham *but* hom, *referring to some kind of hollow space. The added term* strings *refers to the tendon. Others say the whole thing has to do with pigs. Butchers used to hang their ham or pigs by the legs through the space behind the femur, where the tendons behind the knee are located. Either way, this pretty much describes where these tendons are and what they look like.*

The biceps femoris has two separate heads that attach to the ischial tuberosity of the pelvis and the *linea aspera* on the back of the proximal femur. They come together distally to insert together into the head of the fibula. Its main function is to flex the knee, but it will also extend the hip joint. It helps to laterally rotate both the knee and the hip.

The semitendinosus is mainly made of tendon. It starts at the ischial tuberosity in the pelvis and attaches to the medial tibia. It acts to flex the knee, medially rotate the thigh, and extend the hip. The semimembranosus is beneath this muscle. The semimembranosus also starts at the ischial tuberosity and attaches at the *medial tibial condyle*, where it acts to flex the knee, extend the hip joint, and medially rotate the thigh. These muscles are shown in this figure:

132

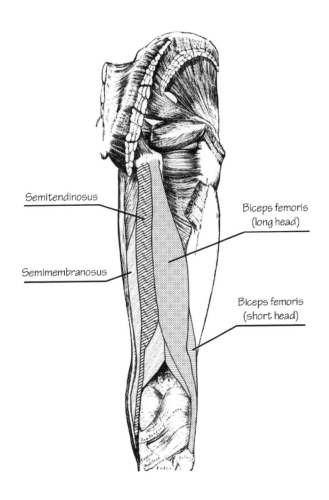

Semitendinosus

Semimembranosus

Biceps femoris
(long head)

Biceps femoris
(short head)

Lower Leg Muscles

The lower leg has muscles that go all around it. The anterior compartment is located on the front of the lower leg below the knee. Because they work on the foot and ankle, these are muscles that will flex, extend, *dorsiflex* (raise up), and invert the ankle and foot. As in the lower arm, there are muscles that also extend the toes—mainly the *extensor hallucis longus* and *extensor hallucis brevis* muscles. There are two other muscles in the anterior compartment besides these two:

- **Tibialis anterior:** This muscle is seen on the lateral surface of the tibia, and is the main muscle that dorsiflexes the foot and ankle. It also inverts the foot, because it ends at the base of the first metatarsal and medial cuneiform bones of the foot. At its proximal end, it attaches to the lateral tibia.
- **Fibularis tertius:** This muscle isn't present in everyone. It starts near the bottom of the *extensor digitorum longus* and attaches to the fifth metatarsal. Its job is to dorsiflex and *evert* the foot.

This is what the muscles of the anterior compartment look like:

133

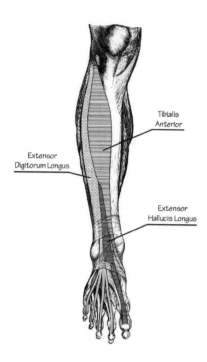

The lateral compartment of the foot has just two muscles. These are called the *peroneus longus* and *peroneus brevis* or the *fibularis longus* and *fibularis brevis*. They act together in order to evert the foot, which is the same as turning the sole of the foot in an outward direction. Most of the time, though, we don't actually evert the ankle or foot, so these muscles just fix the ankle in space while walking or running. This is what these muscles look like:

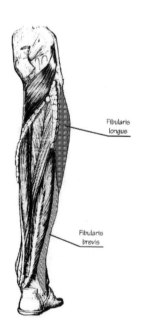

The fibularis longus is more superficial and is located along the lateral part of the lower leg. It starts along the upper and lateral aspect of the fibula, goes behind the *lateral malleolus* of your ankle (the bone that sticks out on the outer ankle) and inserts on the bottom of the foot at the medial cuneiform and base of the first metatarsal. It helps to plantar flex the foot as well as evert it.

Fibularis brevis is shorter than the longus muscle. It starts lower down on the fibula and soon attaches to the same tendon that belongs to the fibularis longus muscle. Its main job then is to evert the foot, because it attaches to the fifth metatarsal base rather than the first metatarsal, which is where the fibularis longus attaches.

The posterior part of the leg has seven different muscles in two layers: superficial and deep. This is a large compartment that serves to plantar flex and invert the foot. The superficial muscles are the ones that form the shape of your calf. All of them insert into the calcaneal tendon, which is your Achilles' tendon on the back of your heel.

The *gastrocnemius* muscle is superficial and has two large heads on either side of the back of the knee. They come together distally to form the main muscle bundle that inserts into the back of the heel at the Achilles' tendon. It is a very strong muscle that mainly plantar flexes the ankle. It also slightly flexes the knee.

Fun Factoid: *What does it mean to have an Achilles' heel? It means to have a fatal flaw or some area of weakness. The whole thing stems from an ancient Greek legend about Achilles, who was a hero and warrior of the time. His mother wanted him to be immortal, so she dunked him headfirst into the River Styx. The only problem is that she didn't dip his heel into the river, which became his weakness and his major downfall in the end. Your own Achilles' tendon is a weak spot that can rupture if you stress it too much.*

The *plantaris* muscle is a minor one—so minor that ten percent of people don't have one. It is long and skinny and located beneath the gastrocnemius muscle. It starts at the distal femur then descends along the medial part of the leg, to end at the Achilles' tendon. It plantar flexes the ankle, but of course it isn't necessary for this process, because the gastrocnemius is so powerful.

The *soleus* is a deeper muscle that is larger and flatter than the gastrocnemius muscle. It starts at the proximal tibia along the *soleal line* and inserts along with the rest of the muscles of this group in the Achilles' tendon. Its job is to assist in plantar flexing the foot. This figure shows what these muscles look like:

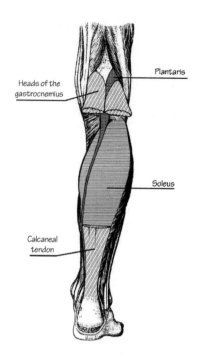

The *deep muscles* are a group of four muscles that act in different ways. The *popliteus* muscle only acts on the knee itself, while the other three act on either the ankle or the foot. This figure shows what these muscles look like:

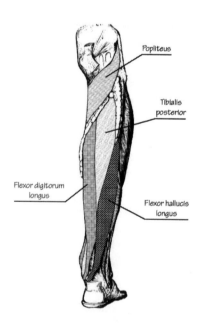

The popliteus muscle is near the knee, forming the base of the space behind the knee called the *popliteal fossa*. It starts on the lateral aspect of the femur and on the back of the lateral meniscus in the knee itself. It then runs downward and medially, so that it can insert into the same place that the soleus muscle starts.

The other three muscles include the *tibialis posterior* that starts between the tibia and fibula in the *interosseus membrane*. It travels down to the foot behind the *medial malleolus* (the inside ankle bone) in order to insert into the medial tarsal bones on the bottom of the foot. Its job is to plantar flex and invert the foot. The medial arch of the foot is formed by this muscle's distal tendons.

The *flexor digitorum longus* and *flexor hallucis longus* have similar jobs. Both of them will flex the toes, but only the flexor hallucis longus tendon will flex the great toe. These muscles start along the tibia and attach to the plantar side of each toe in order to flex them. The flexor digitorum longus only works on the lateral four toes.

Foot Muscles

Just as is seen in the hand, there are intrinsic and extrinsic foot muscles. The foot is not as nimble as the hand, but similar intrinsic muscles are still present. The extrinsic muscles of the foot are the ones we just talked about in the lower leg.

Movement of each toe separately (which is really hard to do) happens because of the intrinsic foot muscles. These are divided into those on the top, or *dorsum*, of the foot and those on the bottom, or the *sole*, of the foot.

On the dorsum of the foot, there are two intrinsic muscles. These are the *extensor digitorum brevis* and the *extensor hallucis brevis*. These are deeper muscles. The extensor digitorum brevis starts on the calcaneus and surrounding areas, and ends on the dorsal side of the proximal phalanx of all toes except the great toe. The extensor hallucis brevis muscle is a bit medial to the extensor digitorum longus muscle, starting in the same general area as the extensor digitorum brevis. It ends at the great toe and allows you to extend it. This figure shows these muscles:

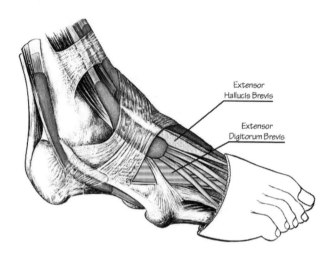

The plantar surface of the foot has ten intrinsic muscles in it. These help to stabilize the arches of the foot and control the toe movements individually. They come in four layers:

- The first layer is most superficial and includes the *abductor hallucis*, the abductor digiti minimi, and the *flexor digitorum brevis*. The abductor hallucis allows you to flex and spread out your great toe. The flexor digitorum brevis will flex the lateral four toes by inserting into middle phalanges of these toes. The abductor digiti minimi abducts and flexes your pinkie toe. This is what they look like:

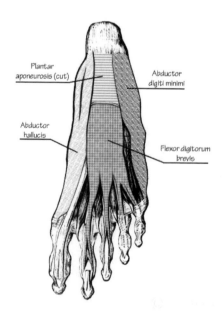

- The second layer has the *quadratus plantae* muscle and the lumbricals. The quadratus plantae will help flex the lateral four toes and itself attaches to the flexor digitorum longus muscle. The lumbricals start where the tendons of the flexor digitorum longus stops. Each of these four muscles sits between the metatarsals of the foot. This is what these muscles look like together:

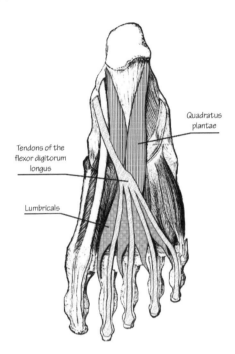

- There are three muscles in the third layer. There is the *flexor hallucis brevis* and *adductor hallucis* muscles, each of which acts to move the great toe in some way. The other muscle is on the other side of the foot, the flexor digiti minimi brevis, which flexes the pinkie toe. This is what these muscles look like:

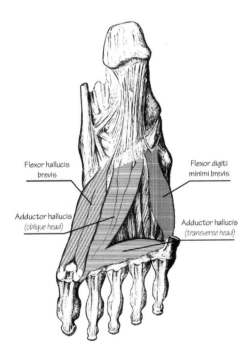

- The deepest or fourth layer is made up of the plantar and dorsal interossei muscles. The plantar interossei muscles adduct the digits and flex them at the metatarsophalangeal joints. The dorsal interossei abduct the lateral toes and help to flex these toes at the metatarsophalangeal joints. This is what they look like:

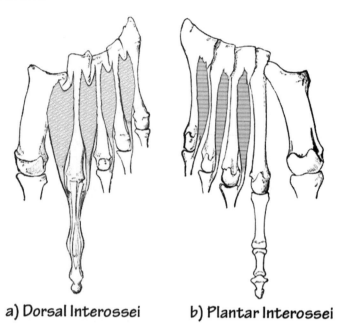

a) Dorsal Interossei   b) Plantar Interossei

To Sum Things Up

That was a lot of muscles to remember, so if you remember the main ones, and the fact that there are groups of muscles that carry out various movements and stabilize the body, you already know quite a bit about how the muscular system interacts with the skeletal system.

Chapter Eight: Your Heart and the Blood It Pumps

In this chapter, you will learn some amazing facts about how blood is pumped from the heart to all parts of the body and back again, in a closed-loop system that runs continuously throughout your life. It's not an easy job for the heart, which does the majority of the work in this system—collectively called your *cardiovascular system*.

Where Is the Heart Located?

In humans, the heart is located in middle of the chest in a space called the *middle mediastinum*. This figure shows you where to find it:

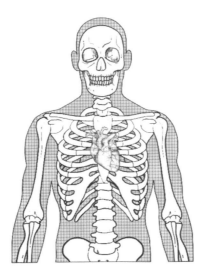

It is enclosed in a protective, double-membraned sac known as the *pericardium* that surrounds the heart and keeps it in its proper place. The back of the heart is located near the vertebral column, and the front of the heart sits behind the rib cartilage and sternum. The upper part of the heart is where several large vessels are attached. These are the superior and inferior *vena cavae*, the *pulmonary vasculature*, and the *aorta*. This part of the heart is called the *base of the heart*, even though it's located at the top. The bottom of the heart is called the *apex*. There are no major vessels coming off of the apex, but this is where the heart muscle is the thickest and strongest.

The largest part of the heart is usually a bit offset to the left side of the chest cavity, which is why a heartbeat can be better heard there. Because the heart is situated between the two lungs, the left lung is smaller than the right lung and has a cardiac notch in it that helps make room for the heart.

The heart is cone-shaped, with the base on the top and the apex at the bottom. The adult heart weighs about 250 to 350 grams. It is about five inches in length, 3.5 inches in width, and about 2.5 inches thick. Well-trained athletes often have a larger heart due to the effects of exercise on the heart muscle.

There are some heart muscle diseases, too, where the heart muscle is abnormally thick. Having an enlarged heart doesn't always mean the muscle is thick. When a person has heart failure, the muscle is often of normal thickness (or even too thin), but the internal chambers are boggy and have too much blood in each chamber.

Fun Factoid: There is a disease called hypertrophic cardiomyopathy *that involves an area of the left ventricle of the heart that is abnormally thickened. Most people have no symptoms until they die of a heart rhythm disturbance, often in adolescence. It is commonly the reason why an apparently healthy athlete will die suddenly of a heart attack. It's a hereditary condition that is* autosomal dominant, *meaning a person will pass it on to half of their offspring. By itself it is rare, affecting fewer than one out of a thousand people, but all close relatives of someone who has it should be tested themselves for the disease.*

How Your Heart Works

Your heart is basically a well-coordinated muscle that beats relatively steadily, at about 60 to 100 times per minute in a healthy adult at rest. The amazing thing about the heart, though, is that it doesn't stay at that rate when you exercise, but has surge capacity to increase both the heart rate and the amount of blood it pumps out per beat. This ensures that your muscles get the vast amounts of oxygenated blood they need for whatever exercise you're doing.

The average person who doesn't exercise much will easily raise their heart rate with exercise. The heart works faster to put out blood, but doesn't necessarily put out huge amounts of blood per beat. Think of the heart like a factory. When demand for the product (oxygenated blood) goes up, the workers in the factory have to work faster in order to keep up with the demand.

This doesn't remain the same as the demand stays up or, in this case, as a person exercises more regularly. When an athlete exercises regularly, there is always an expectation that added demand will be necessary, so the heart compensates for this. If the heart was a factory, it would hire more workers so that more of the product can go out each day, and every factory worker wouldn't have to work so hard individually. The process is more efficient. Increased demand means more output per day, and not just because there is faster output by each worker.

The heart of an athlete is like a factory that has hired more workers. It doesn't have to beat so many times a minute while exercising, because the heart muscle itself is thicker and more efficient. It pumps out more blood each beat so that not as many total beats per minute are necessary. You can see this in elite athletes who often have a resting heart rate of as low as 40 beats per minute.

If you are an average person, though, your heart beats about 80 times per minute, 115,200 times per day, or more than 42 million times per year. If you are lucky enough to make it to the age of 80, your heart would probably have beaten about 3.4 billion times in total. This is why you really need to take care of it.

The amount of blood your heart puts out in one minute is called its *cardiac output*. This is calculated by knowing the number of beats per minute and multiplying that by about 70 milliliters of blood ejected per heartbeat by the heart. This amounts to about 4,900 milliliters per minute, or about 5 liters. When you exercise, this number could go way up so that, if you are doing an extreme sport, your cardiac output could be as high as 30 liters per minute.

Every day, your heart pumps about 7,200 liters (or about 1,900 gallons) of blood. Most adults have about 5 liters of blood inside their entire body, which means you circulate your entire blood supply in just about a minute's time. Added to this feat is the fact that your circulatory system—all of your arteries, capillaries and veins—add up to a total length of about 96,000 kilometers, or about four times the distance around the entire Earth at the equator. Now are you impressed with how great your heart really is? Let's take a look at this remarkable part of your cardiovascular system next.

The Heart's Anatomy Explained

Your heart is a muscle that's about the size of your fist. It is made from cardiac muscle cells that have what's called *autorhythmicity*, which means it doesn't take continued stimulation by any nerve to keep the heart muscle cells happily contracting away no matter what you do. Most of the cells of the heart could probably be called *follower cells*. They carry out the muscle contraction of the heart but don't necessarily lead the way.

Fun Factoid: Your heart rate is largely controlled by your autonomic, or involuntary, nervous system. One branch, called the sympathetic branch, *speeds up the heart, while the other branch, called the* parasympathetic branch, *slows down the heart. If you disconnect both of these branches, your heart would still beat but at a rate of around 100 beats per minute. Thanks to your parasympathetic nervous system, which puts a brake on your heart rate, your resting heart rate is usually much lower than that. A heart rate of about 60 beats per minute in most non-athletes is as low as it will naturally go when this system is fully functional.*

There are few specialized cells in the heart you could call *leader cells*. These are strategically located to start the heartbeat in a coordinated fashion. You don't absolutely need these leader cells, because any cardiac muscle cell can technically contract without help, but it's like having an orchestra play a song without a conductor to guide pacing and rhythm.

The main goal of the heart is to receive deoxygenated blood from the body, which has had much of the oxygen removed by the cells of the body (that use it up during aerobic respiration); pump this blood to the lungs through what's called the *pulmonary circulation*; receive the oxygenated blood back from the lungs; then send it all back out again. The blood is refreshed with oxygen and ready to go through systemic blood circulation all over again.

All this pumping requires coordination and an ability to separate blood without oxygen and blood with oxygen. If these two types of blood mix in any way, it really doesn't help the heart's job, which is to only send out oxygenated blood to the body's cells. This is why the human heart has four separate chambers. Let's look at these chambers and examine the flow of blood through each one.

The Chambers of the Heart

The heart is similar to a shipping center. Items get shipped in, sent for processing, processed, and then shipped out. It all happens in an orderly fashion, with the same number of incoming things getting processed to become outgoing things. In the heart, the four chambers each have a job to do. They take the incoming deoxygenated blood, send it for processing in the lungs, receive the processed oxygenated blood, and ship it back out to the body.

These are the four chambers of the heart:

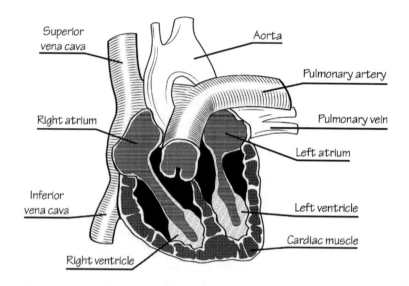

The heart has a left side and a right side. Each side has its own atrium and ventricle. They are arranged in such a way so that blood can be efficiently transferred from one chamber to the next as the heart beats in a coordinated fashion.

Blood first enters the heart from the periphery of the body, from either the superior or inferior vena cava. The inferior vena cava ships blood from the lower half of the body, while the superior vena cava ships blood from the upper half of the body. They come together to dump blood into the right atrium. This blood has had oxygen extracted from it, so it needs to be recharged before it can be sent back out to cells again. There is no valve between the vena cavae and the right atrium. The pressure difference between the two areas draws blood into the right atrium.

The right atrium by itself isn't big enough or strong enough to send blood to the lungs for recharging. It's mainly just a receiving chamber. To actually ship the blood off to be processed by the lungs, it needs a better kick. This is where the right ventricle comes in. There is a valve called the *tricuspid valve* that opens to send blood to the right ventricle, but closes again to keep blood from flowing backward when the right ventricle contracts.

The right ventricle is the main pumping chamber on the right side of the heart. The blood is still deoxygenated because it hasn't gotten recharged with oxygen in the lungs yet. When the heart contracts, the right ventricle (which has a thicker muscle wall than the right atrium) pumps this blood out of the heart to the lungs, where recharging takes place. There is a *pulmonic valve*, or *semilunar valve*, that opens to let blood out of the right ventricle, then closes to prevent backflow into the chamber afterward. Blood leaves the heart through the pulmonary arteries.

We will talk more in a future chapter about the magic that happens in the lungs in order to turn deoxygenated blood into oxygen-rich blood. This blood is perfectly packaged to leave the heart once it gets back to it, and the heart has the muscle power the blood will need to go back out to the body. Before it can do this, however, it needs to go to another receiving chamber called the left atrium. There is no valve between the lungs and this chamber. Oxygenated blood comes into this chamber from the pulmonary veins.

This is probably a good time to talk about what we really mean by arteries and veins. You've probably heard that arteries have oxygenated blood and veins have deoxygenated blood. This is generally true, but it helps to think of these vessels in a slightly different way.

Arteries are any vessel that is traveling outward from the heart, regardless of what kind of blood is in it. This is why the pulmonary arteries leaving the right ventricle are called arteries, even though they are low in oxygen. The pulmonary veins that come from the lungs into the left atrium are still called veins, even though they are high in oxygen.

The left atrium is the receiving chamber for this oxygenated blood from the pulmonary veins. This blood then flows through the *mitral valve* between the left atrium to the left ventricle. Once the blood reaches the left ventricle while this ventricle is relaxed, the mitral valve closes to prevent blood from backing up into the left atrium again.

The left ventricle is the strongest muscle part of the heart, with the thickest muscle walls. It needs to be this strong in order to allow blood to get from the heart to all parts of the body and back to the heart again. As it leaves the left ventricle, the blood travels through the *aortic valve*, which closes after the contraction to keep blood from flowing back into the heart.

Fun Factoid: If the left ventricle is so much thicker and bigger than the right ventricle, does it pump out more blood from the heart with each heartbeat than the right ventricle? This sounds logical but it doesn't work that way. An equal amount must leave each chamber. Otherwise, over time part of the system would have too much blood and part of the system would have too little. If the left ventricle pumped out more blood than the right ventricle, it would quickly shrink the amount of blood left in the lungs. The only way it can work is if the right ventricle puts into the lungs the exact same amount as the left ventricle puts into the body per beat.

If a valve is leaky and blood flows backward when it isn't supposed to, this is called *regurgitation*. Aortic regurgitation and mitral regurgitation can both cause inefficiency of the heart overall. If a valve is narrowed instead so that blood can't get through it freely enough, this is called *stenosis*. Aortic stenosis is a heart valve problem in which the left ventricle pumps blood out, but there is resistance to that blood flow by the narrowed (and often calcified) aortic valve. It is possible for a valve to have stenosis and regurgitation at the same time.

This flow of blood from one part of the heart to the other is called the *cardiac cycle* and involves regular contractions and relaxations of the heart. The atria contract essentially at the same time, and the ventricles also contract at the same time. The ventricular contraction phase is called *systole* and the ventricular relaxation phase is called *diastole*.

When you check your blood pressure, it is a rough measurement of the pressure in the heart during systole and diastole (although the pressure diminishes a little from the heart to your arm where the blood pressure is taken). The top number is the systolic blood pressure and the bottom number is the diastolic blood pressure. The flow of blood through the heart is depicted in this picture:

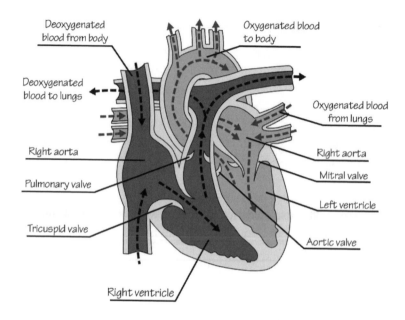

More on the Heart Valves

As we just discussed, the heart has four valves that separate its chambers. There is one valve that separates each atrium and one valve that sits at the exit of each ventricle. For example, there is the tricuspid valve, which sits between the right atrium and the right ventricle. The tricuspid valve has three cusps as well as three *papillary* muscles that project from the wall of the ventricle.

These papillary muscles extend from the heart wall to the valves by means of cartilage-containing connections known as *chordae tendineae*. These muscles are present in order to protect the valves from falling too far back when they close. During the diastolic phase of the heart cycle, the papillary muscles are also relaxed and the tension on the chordae tendineae is less. As the chambers of the heart contract, so do the papillary muscles. This causes tension on the chordae tendineae and helps to hold the cusps of

146

the atrioventricular valves in place, keeping them from sending blood back into the atria. This shows you what these papillary muscles look like:

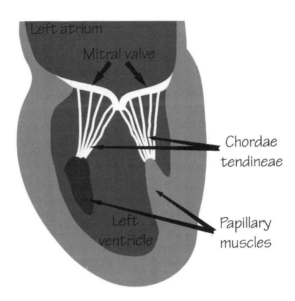

The mitral valve is similar to the tricuspid valve but is located between the left atrium and the left ventricle. Early anatomists called it the mitral valve because it has two cusps that look like a mitre/miter, the hat that bishops wore at the time. It has its own set of chordae tendineae and papillary muscles, just like the tricuspid valve. It closes after diastole to keep the blood from backing up during systole.

Fun Factoid: The mitral valve papillary muscles are particularly vulnerable after a major heart attack. If they get damaged when a heart attack interferes with their circulation, they can suddenly rupture. The end result? Well, about half of all people suddenly die when this happens, which is why the few days after a heart attack can be just as risky as when the heart attack first happens.

There are two additional semilunar valves that sit at the exit of each ventricle. The pulmonary, or pulmonic valve, is located at the base of the pulmonary artery. This has three different cusps that aren't connected to any papillary muscles. When the ventricle is relaxed, blood flows back into it from the pulmonary artery and the blood fills the pocket-like valve, pressing on the cusps to seal the valve.

The semilunar aortic valve is at the opening to the aorta (and the exit point of the left ventricle). It is also not attached to any papillary muscles. It consists of three cusps that close with the pressure of the blood that flows back from the aorta after systole has completed itself. Just outside this valve are some aortic sinuses that open into small coronary arteries and cover the heart while providing it with circulation.

The Pericardium Around the Heart

The heart can be thought of as a grapefruit with a tough peel around it. This tough peel in the heart is called the pericardium. It is a double-layered connective tissue membrane around the heart that serves several important functions.

The pericardium on the outer portion is not very elastic and is called the *fibrous pericardium*. It folds back on itself to cover the surface of the heart as an inner layer. This inner layer is thinner and not as fibrous. It is called the *serous layer*. Between the two layers is a small amount of lubricating fluid that helps to make the heart slippery as it contracts continuously. The serous pericardium has two of its own layers. One is actually stuck to the heart itself, called the *epicardium*. The outer serous layer is called the *outer parietal layer*. It lines the inside of the fibrous layer.

The different roles of the pericardium include the following:

- It keeps the heart fixed in the chest cavity so that it doesn't move around much. It has many attachments to other areas of the *thorax* in order to keep all the structures where they need to be.
- It is tough enough to prevent overfilling of the heart. The heart muscle can expand during diastole, so, in order to keep it from getting too large over time, the pericardium holds the muscle back.
- As mentioned, it is highly lubricating to the heart as it contracts. The serous layer makes this fluid and keeps it at about 50 milliliters or less in the entire chamber.
- It prevents infection from spreading to the heart muscle from the lungs, acting as a barrier to pathogens that might infect both.

Fun Factoid: The 50 milliliters of fluid in the pericardial sac *around the heart is just about right. In some cases, like if a person gets a viral infection around the heart,* cardiac tamponade *can result and challenge the heart's ability to fill with blood during diastole. The blood backs up and the cardiac output plummets, leading to death if the fluid isn't drawn out from between the layers of the sac.*

This image shows what these layers look like along with the layers of the heart:

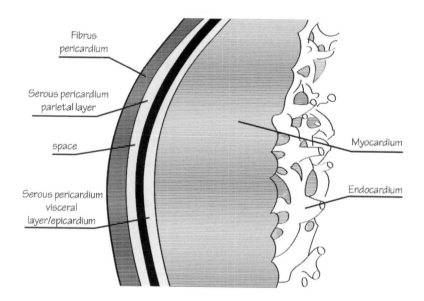

The Heart Wall Layers

There are three distinct layers to the heart wall. These are the epicardium, *myocardium* and *endocardium*.

The endocardium is the innermost layer of the heart. It is made of loose connective tissue covered with *epithelium*. This epithelium is considered structurally similar to the lining of the blood vessels, which is called the *endothelium*. The *endocardial epithelium* is basically a continuation of the endothelium in the blood vessels. Beneath this is the *subendocardial layer* that connects the endothelium to the myocardium further outside the heart wall. Some of the electrical fibers within the heart are located in this space.

The myocardium is a thick muscle layer of the heart wall. This is where the *cardiomyocytes*, or heart muscle fibers, are located. It is of various thicknesses, depending on where in the heart you're looking. It is thickest in the left ventricle and thinnest in the two atrial walls.

There are two types of cardiac muscle cells in the myocardium. There are the muscle cells that contract the heart, and *pacemaker cells* that control the conducting system. The contractile muscle cells, or follower cells, make up 99% of the cells of the heart in the ventricles and atria. They are connected by *intercalated discs* that allow for a rapid response to potential impulses from the pacemaker cells.

The intercalated discs are also present, to allow the cells to act in tandem as a *syncytium*, and enable synchronous contractions that pump blood through the heart and into the major arteries in a coordinated way. This is ideal for the heart, because the muscle cells need to contract as a single unit.

The pacemaker cells of the heart make up about one percent of the heart cells and form the conduction system of the heart. These are the leader cells when it comes to electrical activity. They are generally smaller than the contractile muscle cells, and have very few myofibrils in them so that they don't really contract themselves. Their function is similar in many ways to neurons, or nerve cells. An impulse started at one point in the heart spreads rapidly from cell to cell, in order to trigger the contraction of the whole heart at one time.

The *subepicardial layer* sits between and joins the myocardium and the epicardium. The epicardium is one of the two layers of the serous pericardium. It secretes *pericardial fluid* into the *pericardial space*. It also has fat that allows the coronary arteries that supply the heart to be cushioned and protected as they surround the heart.

An interesting fact about the *coronary circulation*, which is the blood vessel group that supplies oxygenated blood to the heart, is that it travels on the outside of the heart. The arteries form a web around the heart and send branches inward to cross the layers of the heart, from the outside to the inner layers. This means that the inner layer of the heart is more vulnerable to damage in a heart attack than the outer layers. The mildest form of heart attack is called a *subendocardial myocardial infarction* because it only affects the inner layer of the heart.

Another interesting fact about the coronary circulation is that it isn't as well-designed as some other things. While there are plenty of places in the body that have blood supply redundancy, this isn't true of the heart to a great degree. *Redundancy* means that if one blood vessel gets blocked, another blood vessel can take over to supply the same area. When a coronary artery gets blocked, there isn't a lot of this redundancy, so the heart is more vulnerable to damage, even though it's so essential to survival.

How the Cardiac Conduction System Works

Like a real orchestra, the heart muscle cells must act in concert with one another. There must be a consistent source of at least one good leader and an orchestra that follows suit to perform a coordinated pattern of behavior. With these elements, the whole process works with the greatest degree of efficiency, and there aren't any heart muscle cells doing things they shouldn't or that are out of step with the rest of the system.

The heart is allowed to be the most efficient by having certain cells that *depolarize* first. These cells direct the rest of the heart muscle cells to beat along with whatever heart rate the leader cells have decided to set as optimal. The heart has an ingenious system of getting an initial heart rate set and then establishing this rate for the rest of the heart. It does this through its *cardiac conduction system*.

There are specialized conduction cells that don't participate much in contracting the heart, but have the responsibility of initiating a heartbeat and propagating the signal to all parts of the cell. There are four areas where these conduction cells are located in the heart. Each of these areas is connected into a system that starts in the right atrium and travels to the ventricles.

The cardiac conduction system consists of these four main structures: the *sinoatrial* or *SA node*; the *atrioventricular* or *AV node*; the *Bundle of His* or *atrioventricular bundle*; and *Purkinje fibers*. Here's what this system looks like:

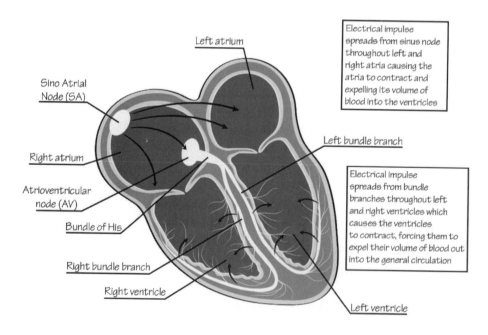

The major leader cells of the heart originate in the sinoatrial node, which is located in the right atrium. It is the primary pacemaker cell because it depolarizes the fastest of all the conducting cells. By depolarizing the fastest, it jumps the gun on other cells and forces them to follow along before they can generate their own pace, which would normally be slower than the beat set by the SA node. When this beat starts, it sends its electrical message across many pathways in the right and left atria, causing them to contract.

Eventually, the signal reaches the AV node, located near the junction of all the cardiac chambers. Here, the cells of this node will have a slight delay. This is enough to allow the atria to contract and fill the ventricular chambers. This will make it more efficient and allow the ventricles to contract after they have been maximally filled with blood. The amount of delay through this node is about 120 milliseconds.

Next, the electrical signal travels down the *septum* between the two ventricles in a group of fibers collectively called the *Bundle of His*. This travels all the way down to the apex of the heart, where it splits into one half that goes to the right ventricle and one half that goes to the left ventricle. These are called *bundle branches*. There's a left bundle branch and a right bundle branch.

Within the ventricles themselves are Purkinje fibers. These allow for the spread of the electrical signal across the entirety of the ventricles. They can contract in a direction that goes from the apex to the base, where the exit valves are located. Unlike the Bundle of His, the Purkinje fibers have no clear path, but instead spread in what is kind of like a wave.

Each of these areas of the conduction system will depolarize and set up its own rhythm at a specific rate that decreases as you go down the conduction pathway. This means that, if the SA node for example does not fire off a signal, there are other parts down the line that will take over and establish a beat, so that the heart will continue contracting. The further down on this pathway the impulse is started, the slower the heart rate will be, because these parts of the conducting system won't depolarize or start up a beat as fast as the SA node.

Because the SA node cells have the fastest intrinsic heart rate, it essentially overrides the other areas and effectively silences their ability to set their own beat. These cells further down the pathway are still considered leader cells but follow the SA node's lead, which is the primary conductor of this orchestra, or heart. This gives rise to the normal resting heart rate of about 60 to 100 beats per minute in the average person.

There is input from your nervous system that helps to direct the heart rate at any given point in time. The part of the nervous system that most controls your heart rate is your autonomic nervous system. The parasympathetic branch is the part that slows your heart rate by sending a signal to the SA node, telling it to slow down its firing rate. The sympathetic branch is the "fight or flight" branch that gets activated with stress, fear or exercise. This increases the firing rate of the SA node so that your heart rate increases as well. This is largely involuntary, although your conscious experiences affect this system greatly.

The Different Types of Circulation in the Body

Your circulatory system is the plumbing of the body. This is where the blood goes after leaving the heart. No part of the body can function properly unless the circulatory system is able to send blood, oxygen and nutrients to it. Most tissues will have a direct connection between the capillaries that contain blood cells and the tissue cells that need the oxygen and nutrients. There are some tissues like cartilage that do not have a direct blood supply, so they rely on the diffusion of essential substances from the actual circulatory vessels to the cells of the tissue.

You might think of circulation as a single circulatory system that takes blood from the heart, moves it around the body, and sends it back to the heart and lungs in order to recharge the blood with oxygen.

There are actually several different types of circulation in the human body. These are, of course, not completely separate from one another, but will be considered separate for the purposes of this discussion.

Pulmonary Circulation

The *pulmonary circulation* is the part of the cardiovascular system that carries deoxygenated blood from the heart to the lungs. The term *pulmonary circulation* is different from the *systemic circulation* we will discuss shortly. Here is a depiction of this system:

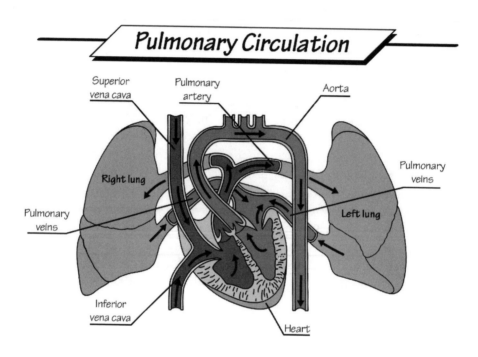

Deoxygenated, or oxygen-poor, blood leaves the right ventricle by means of the pulmonary trunk. Blood is pumped through the pulmonary semilunar valve into the trunk, which breaks off quickly to make the right and left pulmonary arteries.

These pulmonary arteries carry deoxygenated blood to the lungs, where oxygen is picked up during respiration and carbon dioxide is released to air through the lungs' *alveoli*. Large arteries further divide into smaller *arterioles* and finely into extremely small, thin-walled capillaries. This is where oxygen is picked up in the pulmonary circulation. The blood then enters the pulmonary venous system. Ultimately, the larger pulmonic veins are involved in the return of oxygenated blood to the left atrium of the heart. This completes the pulmonary cycle.

The newly oxygenated blood from the lungs will then enter the left atrium, which sends it through the bicuspid valve, also known as the mitral valve or left atrioventricular valve. There, blood is sent into the left ventricle. The left ventricle contracts and sends blood through the aorta to the rest of the body by means of the *systemic circulation*. When the blood is done being used by the tissues, it returns to the right atrium.

Systemic Circulation

The systemic circulation is the part of the cardiovascular system that carries blood that has become oxygenated by the lungs, carrying it away from the heart and into the body. It returns the oxygen-poor blood back to the right side of the heart by means of small venules and larger veins. The blood in this last part of the systemic circulation has been deoxygenated through the process of *cellular metabolism* in the many tissues of the body.

Systemic circulation carries the oxygenated blood from the left ventricle through the aorta and into the smaller arteries, where it finally ends up in the capillaries in the body's tissues. From the capillaries, the deoxygenated blood travels through a system of veins and finally into the right atrium of the heart.

This figure shows an overview of the systemic circulation:

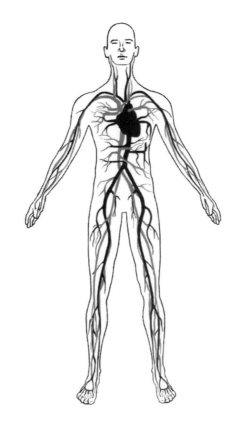

The blood leaves the left ventricle through the aortic valve and goes up into the ascending aorta. The aorta quickly sends off branches, starting with sinuses just outside the aortic valve, that supply blood to the heart itself. Then there are branches to the head and upper extremities, as the aorta takes a U-turn to become the descending aorta.

The descending aorta is long and travels in a downward direction to the pelvis, where it splits off into the left and right iliac artery. As the aorta descends, it sends branches to the lungs, vertebral column, and all the major abdominal viscera. When the left and right iliac artery split off, they send almost all the necessary blood to the left and right legs respectively. As with all parts of this circulation, there are branches that come off all the time to make smaller and smaller arteries to supply every one of the necessary tissues of the body.

Coronary Circulation

The *coronary circulation* is actually part of the systemic circulation, but it's unique in that it supplies the heart muscle itself, so it deserves special mention here. This circulatory system starts with openings in the aorta, just past the aortic valve. Blood exits one opening to make the right coronary artery, which mainly supplies the right side of the heart. It exits another opening to make the left coronary artery, which supplies the left side of the heart. These main arteries have several branches that supply the entire heart muscle. This figure shows the coronary circulation:

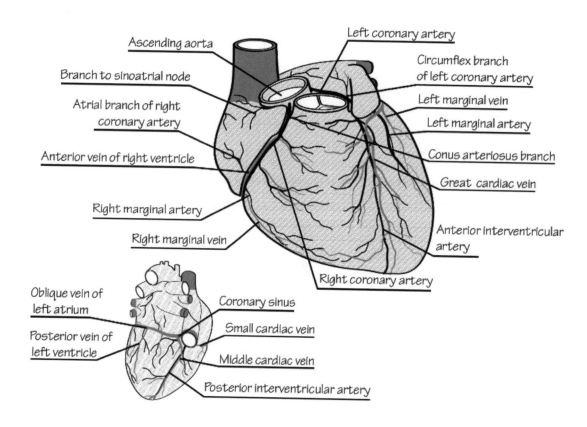

The veins drain back into the right atrium through the coronary sinus. The larger veins of this system generally parallel the track of the arteries, which tend to travel in indentations called *sulci* along the heart surface. The three major veins are the *great cardiac vein*, the *middle cardiac vein*, and the *small cardiac vein*.

Portal System

The *portal system* is a part of the systemic circulation, but has a unique function. This is the part of the circulation that starts at the intestines, where nutrients are picked up in large amounts. The nutrients enter portal veins and then travel to the liver in a short pathway. There are superior and inferior *mesenteric veins* that drain the jejunum through the rectum and a *splenic vein* that drains the stomach, pancreas and spleen. These come together to form the *hepatic portal vein*. This is what drains into the liver.

In the liver, these nutrients are extracted and processed for use by the rest of the body. The portal system effectively ends at the liver, so after processing, the blood then leaves the liver to enter the systemic circulation again.

What Types of Blood Vessels are There?

In this section, we will talk about the roadway system and the process of delivering packages of oxygenated blood from the heart to the cells. This lengthy roadway system involves five different types of vessels that have been uniquely created to do their respective jobs. They all carry blood but, because they need to do different things, their sizes and the walls around these vessels look different, and the vessels themselves behave in unique ways.

Arteries are the vessels that take blood away from the heart. Most of the time this will be oxygenated blood, except when it comes to the pulmonary circulation. These are the vessels that must handle all the pressure exerted upon their walls by the force of the heart during systole, so the walls of arteries must be thick enough. Arterial walls are much thicker than the walls of veins and contain a thick, smooth muscle layer. The smooth muscle contracts in order to maintain the blood pressure as much as possible throughout the long course it takes to get this oxygenated blood to the distal tissues.

There are three major types of arteries. These include the large *conducting arteries* like the aorta and iliac arteries. They are major branches with thick and elastic muscular walls. *Distributing arteries* are those that go directly to a specific organ. The walls are not as thick but these are still muscular arteries. The distributing arteries will branch into the *resistance vessels*, which are small and minimally muscular. These are also called *arterioles*. The pressure in the arterioles is less than the pressure in the conducting arteries.

The walls of each artery have three layers. The inner layer is the *tunica intima*, lined with *endothelial* cells, which are a type of epithelium. The middle layer is muscular and is called the *tunica media*. It is thicker in the conducting arteries than it is in arterioles. The outer layer is made of tougher connective tissue and is called the *tunica externa*. This is what these layers look like:

The Structure of an Artery Wall

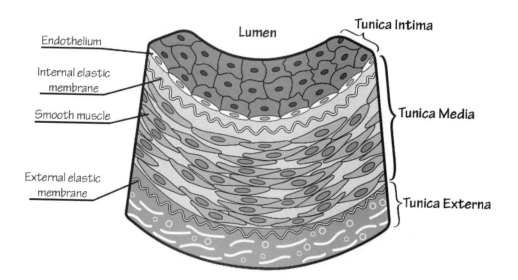

Arterioles are the smaller connecting blood vessels that bridge the gap between the muscular arteries and the capillaries. They have a thin layer of smaller endothelial cells, a thin muscle layer, and a thin tunica externa layer. Arterioles have the job of controlling the flow of blood into the capillaries. This is necessary to keep the pressure in the capillaries from being too high. A thin muscle layer relaxes or contracts in order to regulate the blood flow in these vessels.

Fun Factoid: When you think of heart disease, you might think of atherosclerosis, *which is when cholesterol plaques that contain other substances collect on the inner endothelial layer, narrowing the major arteries. It is much more than just a problem of high cholesterol, though. The cholesterol needs to stick to something. If the endothelial lining is inflamed, damaged, or under high pressure from hypertension, these will be the areas at risk. Other risky areas are where blood vessels divide because the flow is turbulent there. By themselves, plaques can be very stable except if pieces break off and cause blood clots. It's these clots, and not the cholesterol itself, that causes a heart attack.*

Capillaries have extremely thin walls and interface directly with the cells in the tissues. The walls are so thin that they measure just one cell thick, with the ability to have oxygen and carbon dioxide diffuse across the endothelial cells and into or out of the cell. The *lumen*, or the opening of capillaries, is so narrow that in some places, the blood cells must pass just one cell at a time to get through. This is what a capillary looks like:

Capillary

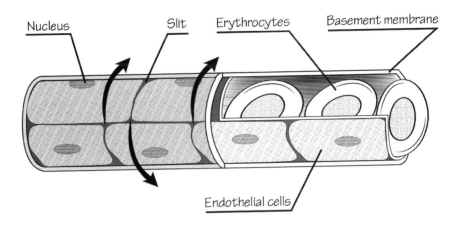

Nucleus Slit Erythrocytes Basement membrane

Endothelial cells

Once the tissues have been recharged by oxygen delivered to their doorstep by the capillaries, the vessels get larger and contain spent blood, which has a higher level of carbon dioxide as a waste product of cellular metabolism and a lower level of oxygen.

The smallest veins are called *venules*, which are usually made when capillaries come together. They have an endothelial cell and thin tunica externa layer, without much of a muscular layer. As the venules travel further out of the tissues and enlarge, the muscular layer increases in thickness and the diameter of the lumen enlarges. At no point, though, does the muscle layer get nearly as thick as it does in the arteries or arterioles.

As venules come together to make veins, the blood flows through these larger vessels, which have thinner walls and larger lumens. They have the same three layers as arteries but these are not nearly as thick as their arterial counterparts. They are elastic and distensible so that blood can fill them up, or they can collapse when there is less blood in them.

Veins have unique projections from their *tunica interna* that form valves in the vessels. The valves are in the veins in order to avoid having the blood flow backward as it travels to the heart. Remember that the arteries have the advantage of the pressure from the heart as it pumps blood through them, and the extra benefit of being muscular enough to maintain that pressure until the blood reaches the capillaries. Here, the pressures must be very low in the tissues themselves.

Fun Factoid: Why do we get varicose veins and not varicose arteries? Part of the problem is the thin-walled nature of the veins that bulge out. Once the vein bulges too much, the valves also fail, and gravity causes the blood to pool in the legs, which is where varicose veins are usually located. If you exercise your leg muscles to push the blood through the veins or wear compression stockings, your chances of varicose veins drop dramatically.

Veins must start out at near-zero pressure in the capillaries while still getting the blood back to the heart. Valves help this process, as well as the external pressure you give to your veins by the tension of your extremity muscles. This forces the blood to move upward (against gravity in some situations) and back to the heart. The muscle layer in the veins don't do much to help this process. This image shows you the difference between arteries and veins:

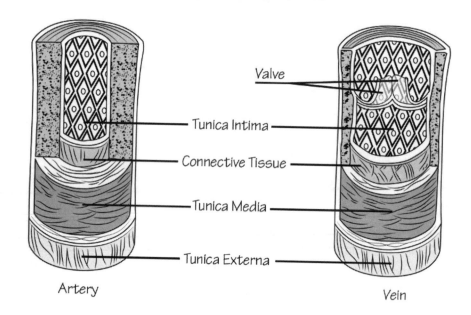

Blood and What Makes It

The hematological system, or blood system, is made from the blood and the organ systems that contribute to the making and/or processing of this fluid. Remember that blood is considered a connective tissue, even though it's basically a liquid. Its job is to contain blood cells, nutrients, waste products and metabolic gases, in order to supply the body's tissues with what they need to function.

While blood looks just like red liquid, it is actually a mixture of liquid and solid parts. If you spun blood in a centrifuge to separate out its components, this is what you'd see afterward:

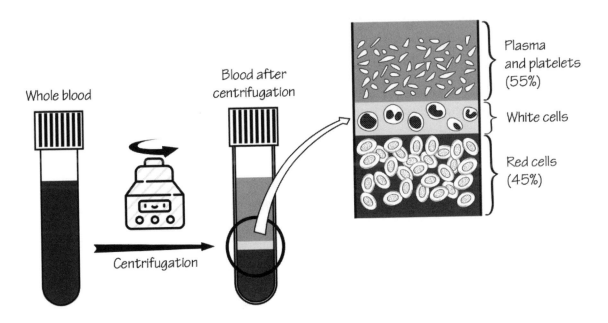

Whole blood

Blood after centrifugation

Centrifugation

Plasma and platelets (55%)

White cells

Red cells (45%)

You'd see that the bottom solid layer of red blood cells represents about 45 percent of the total thickness of the column of fluid in the test tube. A thin layer, called the *buffy coat*, sits on top of that, which is made from white blood cells. The top liquid layer is called *plasma* and is made up of plasma and the tiny *platelets* in them. If you remove the platelets and clotting factors using chemicals, you'd get a completely liquid portion called *serum*.

Erythrocytes, or Red Blood Cells

Starting at the bottom layer are the red blood cells, or *erythrocytes*, which look like flattened discs with an indentation in the middle. They are only about 6-8 micrometers in diameter but are easy to see under a microscope. They are colored red because of the presence of *hemoglobin*, a large protein molecule where a great deal of the oxygen is held. Hemoglobin with oxygen molecules attached is bright red, while hemoglobin without as much oxygen attached is maroon in color. This is why venous blood is not as brightly colored as arterial blood. Red blood cells look like this when they are normal:

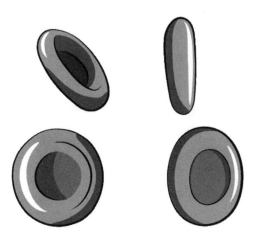

We will talk about where blood cells are made in a minute; but once made, they have a lifespan in the circulation of about 120 days. Those that are old or damaged in some way get taken up by cells and destroyed in the spleen and liver. The iron is recycled to make new red blood cells.

Red blood cells look the same but have different cell *antigens*, or surface proteins, on the outside, depending on your genetic makeup. If you've heard of the term *blood type*, you might already understand part of how this works. Your blood type defines which antigens are located on the red blood cell surfaces. You receive half of these antigen types from your mother and half from your father.

The ABO blood-typing system involves antigens labeled A, B and O. The combination of these leads to several blood types, including AB, B, A, and O. People with type A or B blood can have two A or B components—one from each parent—or can have one parent with type O blood and one with an A or B antigen on their red blood cells. The O is sort of silent and doesn't affect the actual blood type a person has. This is an image of what these cells would look like if you could see the antigens directly:

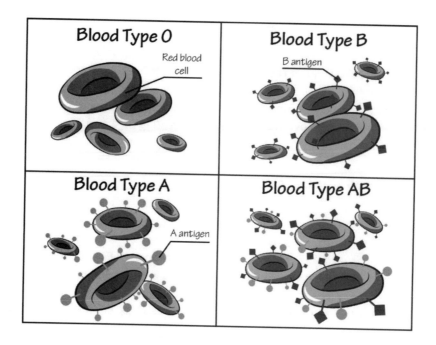

The A and B antigens are kind of picky because, if you have them and give your blood to anyone who doesn't have these antigens in their blood, that person will automatically reject the blood. People with type O blood don't have a lot of immunological reactivity linked to them. Anyone can receive this type of blood. On the other hand, these same people can only receive type O blood themselves because their immune system will reject any other type.

Another red blood cell antigen is called the *Rh factor antigen*, or the *D antigen*. This is an important antigen when it comes to maternal-fetal health because women who are Rh factor negative can be sensitized to their baby's red blood cells if the baby is Rh-positive, and might actively reject the fetus in the womb, sometimes leading to fetal death.

Type AB positive blood is called "universal recipient blood" because people with this type can receive any other blood type in an emergency. Type O negative is the universal donor. They can give blood to anyone.

Leukocytes, or White Blood Cells

There are far fewer white blood cells, or WBCs, in your blood, but these are nevertheless very important for your health. These are the cells that help in various ways to support your immune system functions. A normal person has about 4,000 to 10,000 white blood cells per microliter in their whole blood. There are five different white blood cells you should know about. Each is slightly different with regard to what they look like and what they can do. Let's look at each of these types:

- **Neutrophils:** These are a type of *granulocyte*. They are called *granulocytes* because there are granules under the microscope that you can see with proper staining techniques. They are very abundant—the most prevalent type of WBC in the bloodstream. Neutrophils have multi-lobed nuclei and are responsible for fighting pathogens. Many will go into the tissues directly in order to fight an infection, by engulfing them and secreting substances that help in the marking and destruction of pathogenic viruses and bacteria. They only last 4-7 hours in the bloodstream.
- **Eosinophils:** These are fewer in number than neutrophils and have large bluish granules. These cells have a prominent bi-lobed nucleus. They are active in worm or parasite infections and in the allergic response. They release chemicals that trigger allergic reactions; their numbers are highest in parasitic infections or in those with active allergies. They can live for a few days in the tissues.
- **Basophils:** There are very few basophils in the systemic circulation, representing less than one percent of the total. These cells are also granulocytes with large red granules. They are elevated when inflammation or hay fever is present, and are the cells in the circulation that release chemicals like histamine as part of the immune response. Histamine is especially noticeable when it is released into the bloodstream, because it causes the itching you experience if you have hives.
- **Monocytes:** Monocytes are one of two types of *agranulocytes*, which means they don't have visible granules in them when seen under the microscope. It is also called a *mononuclear cell* because the single nucleus has no specific lobes. Monocytes are large cells representing 2 to 10 percent of all leukocytes. These are major *phagocytic cells*, meaning they can change when pathogens are present in order to eat these foreign agents. They are only called *monocytes* when in the circulation; in the tissues, they are referred to as *macrophages*.
- **Lymphocytes:** These are fairly abundant agranulocytes, representing about 20 to 30 percent of the total number of leukocytes. They will exit and enter the tissues whenever necessary, and last for different lifespans. These are very important in both making antibodies, helping the immune response in general, destroying infected cells, and acting as memory cells that remember past infections. There are *B lymphocytes* that mainly make antibodies and *T lymphocytes* that mainly destroy infected or cancerous cells in the body.

This is what these different leukocyte types look like:

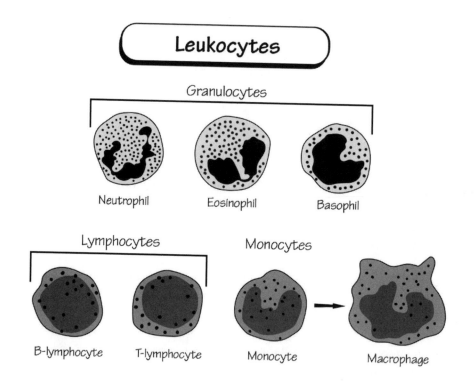

Leukocytes

Granulocytes

Neutrophil   Eosinophil   Basophil

Lymphocytes   Monocytes

B-lymphocyte   T-lymphocyte   Monocyte   Macrophage

Thrombocytes, or Platelets

Platelets aren't really cells, but more like fragments of cells. There are no nuclei inside and they're very small. In your bloodstream, you have about 150,000 to 450,000 platelets per microliter. These are the blood components that are involved in the blood clotting process.

When a blood vessel is injured or broken open, there are a number of clotting factors in the bloodstream that activate the platelets, essentially making them sticky. They stick to the damaged or inflamed parts of the blood vessel and to each other to form a kind of platelet bandage that covers the area in order to prevent further bleeding. Red blood cells also stick to the platelets to form a blood clot. They stay there for however long it takes to fill in the gaps with cells that are more permanent, before the entire clot is degraded and replaced with collagen or scar tissue. The total lifespan of a platelet is about 10 days in the bloodstream.

Platelets actually break off of much larger cells called *megakaryocytes*. Megakaryocytes are huge, multinucleated cells made in the bone marrow. As platelets are needed, they break off piece by piece from the megakaryocyte cell. This is what a megakaryocyte looks like:

The Act of Making Blood: Hematopoiesis

Hematopoiesis comes from the Greek word *poiesis*, which means "to make." It is basically what has to happen in order to make the different cellular components of blood. All blood cell types are made from hematopoietic stem cells, which are basically very immature baby cells in the bone marrow with a lot of potential. By themselves, they have no real function except that they have the ability to divide rapidly and differentiate into any cell in the bloodstream. All the blood components come from the same basic hematopoietic stem cell, or *progenitor cell*.

You can find these stem cells in the *medulla*, or bone marrow, in the bones. This area is basically a blood component-making factory. The stem cells, or HSCs, will divide and differentiate at the same time. But if this is all each cell did, you would eventually run out of stem cells. For this reason, some HSCs stay as a kind of employee float pool, where they remain immature and have the unending ability to divide. Upon cell division, one daughter cell remains immature while the other will differentiate. This is why you never run out of stem cells.

The whole process is asymmetric, meaning that the daughter cells created when an HSC divides has several options. It can remain an HSC and be a part of the float pool of HSCs that keep hematopoiesis going throughout your lifetime. They can also give rise to two different types of progenitor cells, called *myeloid progenitor cells* and *lymphoid progenitor cells*. These are still immature but have started the process of growing up to be a mature cell.

The myeloid progenitor cells have a wide range of possible outcomes, including those that turn into cells not found in the bloodstream. The blood cells that come from myeloid progenitor cells are red blood cells, platelets, monocytes, and all types of granulocytes. These same myeloid progenitor cells can differentiate into mast cells, osteoclasts (that break down bone), and *dendritic cells* (which are tissue-resident immune cells that participate in immune functioning).

The lymphoid progenitor cells are also called *lymphoblasts*. These will further differentiate into any one of these cells: some dendritic cells, T lymphocytes, B lymphocytes, and *NK*, or *natural killer cells*. As you will learn when we talk about the immune system, the T lymphocytes are not considered mature until they have passed through the *thymus* after their release from the bone marrow.

This figure shows how this all breaks down in the end:

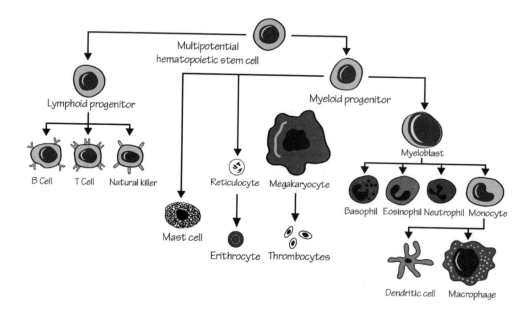

Most but not all hematopoiesis happens in the bone marrow. In the human embryo there are no bones, so the process happens in the yolk sac in places called *blood islands*. As the different organ systems develop, hematopoiesis happens in the lymph nodes, liver, spleen and ultimately the bone marrow when these structures eventually form in the fetus. Over time, the bone marrow takes over the lion's share of the job of making new blood cells.

Maturation of some cells takes place outside the bone marrow. Most red blood cells start out as *erythroid precursor cells* that have nuclei. As they mature, they lose their nuclei altogether. The stage before the mature red blood cell is a cell called a *reticulocyte*. Some of these reticulocytes will escape from the bone marrow before fully maturing in the bloodstream, especially if there is a sudden need for extra red blood cells.

The B cells mostly mature in the bone marrow but do not become fully mature, antibody-making cells until they have been fully processed in the lymph nodes. T cells are also not mature until they have undergone a similar sensitizing process in the thymus gland in the chest cavity. The vast majority of B cells and T cells do not make the final cut, and are destroyed in these organs before ever getting released into the circulation.

In the embryo, fetus, and in some cases the bone marrow when it's not functioning properly, the hematopoietic process must happen in places outside the bone marrow. This is called *extramedullary hematopoiesis*. Sites for this process include the spleen, liver and thymus. All these organs tend to be larger in fetuses, because they need to make blood cells along with their other functions.

Fun Factoid: Some people are unlucky enough to have their own immune system attack their hematopoietic stem cells, so that the float pool of precursor cells dries up. This is very bad and is called aplastic anemia. It means your bone marrow won't make the cells it needs for your blood or immune function. The survival rate is only about 30 percent after a year. Chemotherapy and radiation treatments can also cause this problem, but these will be temporary conditions with a better outlook. In some instances, a bone marrow transplant may be offered as a cure for those with this condition.

There are certain growth factors in the system that tell the bone marrow to make or not make different types of cells. In the case of erythrocytes, or red blood cells, a hormone called *erythropoietin* is made by the kidneys as a growth factor hormone used to trigger red blood cell formation. There are similar growth factors, such as *granulocyte stimulating factor* (also called *colony-stimulating factor*) that turns on the production of granulocytes in the bone marrow.

To Sum Things Up

Your cardiovascular system consists of the heart as the major blood pump, plus the thousands of miles of blood vessels inside your body. Blood must get to every living cell for your body to survive. The heart is an amazing organ that both receives and sends blood to where it's needed in an efficient and coordinated process.

Blood vessels carry blood where it needs to go. The three main subtypes of cellular blood components are the red blood cells, white blood cells, and platelets. They participate in carrying oxygen, fighting the immune system and clotting blood, respectively. All these components arise from a single hematopoietic stem cell found in the bone marrow.

In the next chapter we'll talk about the lungs, as these are important structures to the process of recharging the blood, so that it can both carry oxygen to the cells and be relieved of metabolic waste products. As you'll see, the heart and lungs work closely together in order to participate in the necessary ability of your blood to nourish every one of your living cells.

Chapter Nine: The Respiratory System Explained

You might think that the respiratory system is only about the lungs, and that its only job is to control the exchange of oxygen and carbon dioxide in the body. While this is somewhat true, the lungs and respiratory system have other jobs that are necessary for your health and life. It involves an entire conducting system where air is drawn into the body, carried to the millions of tiny alveoli within the lung tissue, and exhaled again after gas exchange has occurred.

The lungs work hand-in-hand with the circulatory system in order to have efficient gas exchange. Think of the lungs and circulation as a type of purchase at a retail store, in which oxygen is the merchandise given to the circulatory system and carbon dioxide is given back in exchange into the airways.

On the supply side, you need delivery systems and a place to display the merchandise. If the source doesn't have any merchandise for the shopkeeper, nothing will get sold. This would be the case if there wasn't enough oxygen in the environment, which is what happens in high-altitude situations.

Fun Factoid: Altitude sickness is a complex disease with a major symptom called pulmonary edema. The blood vessels in the lungs sense that there isn't enough oxygen around, so they clamp down and constrict. This raises the blood pressure in these arteries, and fluid leaks out of them under such high pressures. Fluid in the tissues that have leaked out like this is called pulmonary edema, which makes it even harder to get oxygen and carbon dioxide exchanged in the lungs. Some people die from this if they don't acclimate to the high altitudes first.

If the delivery trucks are not functioning or the highways are blocked, there won't be any merchandise for sale. This can happen if you can't get a normal breath or stop breathing for any reason. The same would be true if your *bronchial tree* is clogged with inflammatory cells or infection, or if the bronchial tree is so spasmed that air can't get into the lungs. The oxygen is available in the atmosphere but the supply chain that gets it to the alveoli of the lungs isn't working.

If the merchandise was never put out for display, no one would be able to buy it. In the lungs or the alveoli are damaged so that air cannot cross through the respiratory membrane into the circulation, the same issue would occur. There would be plenty of oxygen but no ability to buy it.

Fun Factoid: There are a lot of lung diseases that mess with the alveoli. These tiny sacs in the lungs must be able to take in air. They have very thin membranes so that gases can be exchanged. In diseases like the pulmonary fibrosis of COPD (chronic obstructive pulmonary disease), the alveoli are damaged or are so filled with scar tissue that any oxygen that gets into them has no way of being able to cross through the respiratory membrane in order to get into the blood supply.

Finally, the demand must be there and customers must come, or it won't matter how much merchandise you have. In the lungs, if there is a mismatch so that there are parts of the lungs that have oxygen but the circulation never arrives there, oxygen will still not get into the body. It has no place to go. Blood will pass through areas that have no oxygen in other cases and will come away empty, just as oxygen-poor as it was when it entered the lungs. So, this can get very complicated and depends on all of the parts working effectively together in order to have normal gas exchange in the lungs.

Besides this important gas exchange function, the lungs are partly responsible for controlling the acid-base balance in the bloodstream. The body has just a few mechanisms for controlling the acidity of the blood, and the way carbon dioxide is managed by the lungs is part of how this process takes place. We will talk about this at the end of the chapter.

The Upper Respiratory Tract

There are a couple of ways to divide up the respiratory tract. One way is to divide it into the *conducting zone* and the *respiratory zone.* The conducting zone is the delivery system in this process, while the respiratory zone is where the exchange of merchandise or gases occurs. In the respiratory system, only the alveoli are considered the respiratory zone, while the entirety of the rest of the respiratory system is strictly for conduction of air through the nose, throat and bronchial tree.

The other way to divide up the respiratory tract is to separate it into the *lower airways* and the *upper airways.* The lower airways are also called the *lower respiratory tract* and include the part of the larynx below the vocal cords, the trachea, the bronchi, and the bronchioles. The upper airway, or *upper respiratory tract*, includes the nasal passages and the nose, the vocal cords, the part of the larynx above the vocal cords, the pharynx, and the paranasal sinuses.

The upper respiratory tract is visible as the part of the respiratory system that lies outside of the thorax and above the vocal cords or cricoid cartilage. The larynx is sometimes included in both the upper and lower airways. It all starts at the nasal passages, where you breathe air in through your nostrils. While you can also breathe in through your mouth, most sources do not really talk about the mouth as being part of this system.

The Nostrils

The nose is not the most important part of the respiratory tract, but it does have a role to play. When you take in a breath, you take air into your nostrils and up into the nasal passages. These passages are lined with respiratory epithelium that has cells in it called *goblet cells*. Goblet cells are found in other areas of the body as well. Their only job is to make mucus. There are also hairs in your nasal passageways that trap debris and dust so you don't inhale it.

Inside the nasal passages are some funky-looking ridges called *turbinates*. They increase the surface area of the nasal *mucosa* so that there is more area with which to warm the air as it enters the nasal passages and into the lungs. The goal is to have warm and humid air entering the lungs as much as possible. The turbinates also have pathogen-trapping mucus, so the air is as free of damaging organisms as it can be before it descends into the lungs. If you don't blow your nose, you swallow this mucus and the pathogens are destroyed by your stomach acid. This is what your nasal passages look like:

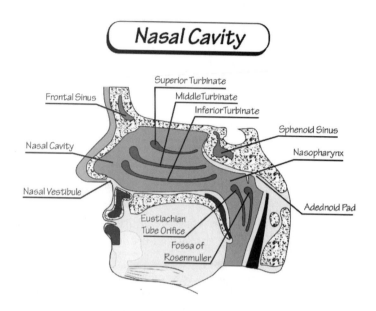

Why Do We have Sinuses?

The sinuses are a connected system of hollow cavities located within the skull itself. In your cheekbones are your *maxillary sinuses*, which are the largest sinuses in the body. In the lower part of the center of your forehead, there are the *frontal sinuses*. Between your eyes are the *ethmoid sinuses* and in the bones behind the nose are the *sphenoid sinuses*. Each of these sinuses are lined with respiratory mucosa that secretes a small amount of mucus for the trapping of pathogens and debris.

The real function of the sinuses, though, doesn't have much to do with the respiratory tract. You see, the skull by itself would be much too heavy without the sinuses to lighten the weight. Instead of solid bone, the sinuses are filled mostly with air, which means we don't have to work so hard to keep our heads upright, and can do so without getting a neckache while we're sitting. Smart idea, no?

The Upper Airways

The air that enters the nasal cavity must travel through parts of the nose, mouth and throat that are designed for eating and drinking as well as for breathing. This requires some coordination so that air goes where it belongs in the lungs, while food instead travels down the *esophagus*.

In order to see how this works, you need to understand a little bit about the anatomy of this area of the body. The main structure involved is called the *pharynx*. This is what we call the *throat*, even though some of this structure is not actually part of your throat. The anatomical definition of the pharynx is that it is the space behind your nose and mouth that extends down to the larynx, or voice box.

There are three separate parts to the pharynx. Each has a slightly different function. The *nasopharynx* is behind the nasal cavity and is considered part of the respiratory system but not the digestive system. It contains your *adenoids*, which are also called the *nasopharyngeal tonsils*. This is a structure of the immune system that stands at the ready where it is most likely that pathogens like bacteria and viruses can enter the body. The part beyond the soft palate, or the roof of your mouth, is the *oropharynx*.

Because it is unlikely that the nasopharynx will be roughened up by the swallowing of food, the lining of this structure is called *respiratory epithelium*. These are columnar cells in a single layer that have a lot of intermixed mucus-producing goblet cells. This is a delicate epithelial layer that would easily get damaged if food would regularly come in contact with it.

The *oropharynx* is below the nasopharynx and is the part behind the mouth. It is a dual-duty structure for both the digestive system and the respiratory system. Because food could disrupt the epithelium, this layer is made of stratified *squamous epithelial cells* that can resist friction caused by food. There are not too many goblet cells in this epithelium.

In front of the oropharynx is the *oropharyngeal isthmus*, which is the opening to the mouth. The *palatine tonsils* are on either side of this structure. Like the adenoids, these are like sentinels ready to fight pathogens where they are most likely to enter the body.

The *laryngopharynx* is the lowest part of the pharynx. It can also be called the *hypopharynx*. It starts at the *epiglottis* and ends at the cricoid cartilage, where the opening of the esophagus is located posteriorly and the opening of the trachea is located anteriorly. Food and air pass through this cavity. There is a recess called the *piriform recess* where food gets stuck most often. There is one of these recesses on either side of the opening of the larynx.

The epiglottis is a big, spoon-shaped piece of elastic cartilage involved in keeping food out of the trachea. When you swallow, the larynx and pharyngeal tissues rise. The rising of the pharynx causes it to open in order to receive food and drink. When the larynx rises, it causes the epiglottis to close off the tracheal opening, forming a lid over the respiratory tract so that food doesn't go down the wrong way.

Label the pharynx below.

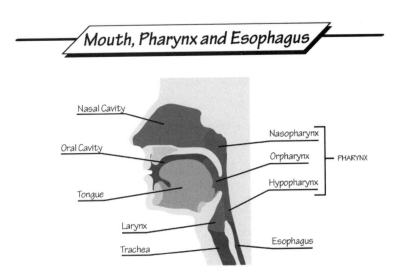

The Larynx, or Voice Box

The voice box itself isn't important to actual breathing, but is still an integral part of the respiratory system. Without the control of your breathing and the way your breath coordinates with your vocal cords, you would not be able to communicate. Just like everything in the body, coordination is everything. You need the different aspects of breathing and vocal cord musculature to say what you need to say at any point in time.

The muscles of the larynx are separated into extrinsic and intrinsic muscles. The intrinsic muscles are both *phonatory* (speaking) muscles and respiratory muscles. The respiratory muscles are tiny muscles that move the vocal cords in order for you to breathe. The phonatory muscles pull the vocal cords together and allow you to make sound. The extrinsic muscles pass between the larynx and surrounding tissue, encasing the intrinsic muscles entirely. There are several of these muscles that must work together to open, close and strengthen your vocal cords, so you can get air out in specific ways in order to make unique sounds.

The extrinsic laryngeal muscles position and support the larynx within the trachea. They include a number of muscles, such as the *genioglossus, hyoglossus, geniohyoid, mylohyoid, stylohyoid, digastric, inferior constrictor, sternohyoid, omohyoid* and *sternothyroid* muscles. These are a lot of tiny muscles that have specific functions. They either depress or elevate the larynx externally. It's a lot of muscles that work in concert with one another to make sure you breathe properly and at the right rate in order to speak.

So basically, the intrinsic muscles are the inner working muscles of the vocal cords that direct just the vocal cords itself. Pitch and tone depend so much on how far apart the vocal cords are from one another and how tense they are. The extrinsic muscles are mainly supportive muscles for the larynx. They work with your major breathing muscles in order to control airflow through the vocal cords. If these don't work, you might find yourself shouting when you really mean to whisper.

This is what your vocal cords and the surrounding cartilage look like:

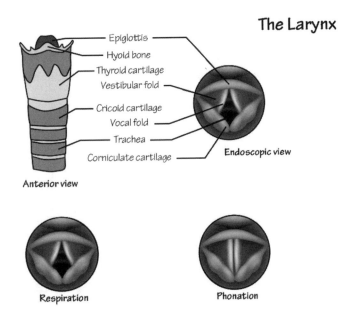

The Larynx

Epiglottis
Hyoid bone
Thyroid cartilage
Vestibular fold
Cricoid cartilage
Vocal fold
Trachea
Corniculate cartilage

Endoscopic view

Anterior view

Respiration

Phonation

The Lungs as Major Respiratory Organs

The lungs are the major structures of respiration and ventilation in human beings. There are two lungs located on either side of the heart near the backbone. Their main job is to extract oxygen from the air and send it to the bloodstream. They exchange oxygen molecules for carbon dioxide molecules that are given off during expiration. The act of ventilation is driven by muscles, particularly the diaphragm and accessory muscles, that contract to allow for the inspiration of air. The expiration of air is usually a passive thing involving no muscular activity, although forced expiration is possible using skeletal muscle action.

The lungs together weigh about 3 pounds, with the right lung being heavier. The lungs are part of the lower respiratory tract that starts at the trachea, branches into bronchi and bronchioles, and finally terminate in the respiratory alveoli. The lungs contain about 1,500 miles of airways and 600 to 700 million alveoli.

The lungs are enclosed by a pleural sac that allows the outer and inner walls to slide over each other in the process of breathing without any friction. This sac encases each lung and also divides the lung into sections, known as *lobes*. The right lung consists of three lobes and the left lung has two lobes. Each lobe is separated into *lobules* and even smaller *bronchopulmonary segments*. The lungs have an interesting blood supply. There is systemic circulation that supplies the *parenchyma* as well as pulmonary circulation that results in air exchange. The systemic circulation is oxygenated and is also known as *bronchial circulation*. This image shows the lungs in the chest cavity:

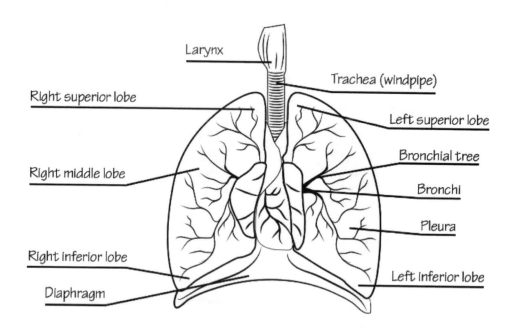

Anatomy of Human Lungs

Larynx

Trachea (windpipe)

Right superior lobe

Left superior lobe

Bronchial tree

Right middle lobe

Bronchi

Pleura

Right inferior lobe

Diaphragm

Left inferior lobe

The Pleura and the Pleural Cavities

The pleural cavity is the thin space between the *visceral* and *parietal pleurae*. It really isn't much of a space because the inner and outer aspects of the pleura are essentially stuck to one another. Between the two layers is fluid that allows for friction-free movement of the lungs, similar to the way that pericardial fluid in the heart lubricates the pericardial cavity.

The pleural membranes are actually just a single layer that folds back on itself in order to form a *visceral* and *parietal layer*. The visceral layer covers the lungs and related structures, including the nerves, bronchi and blood vessels, while the pleural layer lines the chest wall. The pleural cavity can be seen as a *potential* space (meaning it isn't a real space) because normally the two pleural layers are in direct contact with one another.

The pleural cavity, with its pleural membranes, optimizes the breathing process. The surface tension of the pleural fluid adds to the ability of the two pleural layers to be in contact with one another. Together, the layers create a negative pressure environment that allows external air to enter the lungs. We will talk more about how this happens in a little bit.

The Thoracic Cavity

The thoracic cavity, or chest cavity, is the part of your body protected by the rib cage in the thoracic wall. The central compartment of the thoracic cavity is known as the *mediastinum*. There are two entrances to the thoracic cavity—a *thoracic inlet* and a *thoracic outlet*. The thoracic cavity includes parts of the cardiovascular system as well as the lungs—mainly the heart and major vessels.

There are other structures in the thoracic cavity besides those of the respiratory system (the diaphragm, trachea, bronchi and lungs) and the heart and great vessels. These include structures of the digestive system like the esophagus, and miscellaneous organs like the endocrine glands (the thymus) and *lymphatics* (including the thoracic duct). It also contains structures of the nervous system, which include the paired vagus nerves and two sympathetic nervous system chains.

There are three potential spaces inside the thoracic cavity. These include the two pleural cavities (one on each side of the body) and the pericardial cavity. These are lined with mesothelium. The mediastinum makes up those organs located in the center of the chest between the two lungs.

The Mediastinum

The mediastinum is the main central compartment of the thoracic cavity. It is surrounded by loose connective tissue and contains multiple structures within the thorax, including the heart, the major vessels, the esophagus, the trachea, the thoracic duct, the cardiac and *phrenic nerves*, the thymus, and the lymph nodes of the central part of the chest cavity.

The mediastinum is enclosed on the right and left side by the pleural lining. It is surrounded by the lungs on each side, the sternum in the front, and the spine in the back. It contains every organ in the thorax with the exception of the lungs. This is what the heart looks like situated in the mediastinum:

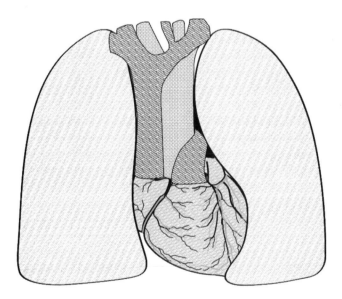

The mediastinum has an upper and lower part. The superior mediastinum begins at the *superior thoracic aperture* and ends at the thoracic plane. The thoracic plane is a totally arbitrary plane that separates the inferior and superior mediastinum. The inferior mediastinum starts at the thoracic plane and ends at the level of the diaphragm.

The Thoracic Cage, or Rib Cage

The thoracic cage is also referred to as the *rib cage* and makes up the thoracic wall. It consists of twelve thoracic vertebrae, twelve pairs of ribs, the sternum, and costal cartilages. It is dome-shaped (like a bird cage) with horizontal bars made up of the ribs and thoracic vertebrae.

The function of the thoracic cage is to support the thorax and to protect the thoracic and abdominal internal organs from being injured. It also resists the negative air pressures made by the elastic recoil of the lungs, provides an attachment for and supports the weight of the upper limbs, and provides an anchoring attachment of many muscles that are involved in respiration.

The Ribs

The ribs are long, curved bones that collectively form the rib cage. In humans, there are 12 pairs of ribs for a total of 24 ribs. The first seven sets of ribs are known as *true ribs* or *vertebrosternal ribs*. They start at the vertebrae and end anteriorly at the sternum by means of costal cartilage.

Rib 1 is harder and of a different shape than the rest of the ribs. After the first 7 true ribs, there are 5 pairs of false ribs. Three of these pairs connect to each other by means of cartilage but not directly to the sternum, while the two bottom-most pairs are called *floating ribs* because they don't connect with anything anteriorly. This is what your ribs look like all together:

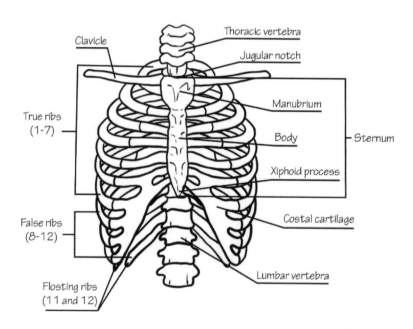

The rib cage is separate from the abdomen by the diaphragm, which is the main breathing muscle for ventilation. When the diaphragm contracts, it causes expansion of the thoracic cavity, reducing the air pressure in the thorax and allowing air to enter. The *intercostal muscles* between the ribs can also participate in the breathing process.

The Lower Respiratory Tract

The lower respiratory tract involves everything below the level of the larynx and includes the trachea, bronchi, bronchioles, lungs and alveoli. Most of this part of the respiratory tract is still like a highway or a conducting zone for the system. These are the parts that deliver air and take away gases exchanged from the alveoli, but do not participate in actual respiration, which is much more specific and happens only in the respiratory zone of the lungs. Let's look at all of the parts of the lower respiratory tract.

The Trachea and Bronchi

The trachea is also known as the *windpipe*. It is a tube consisting mainly of cartilage and connective tissue that connects the larynx and pharynx to the lungs, allowing for the passage of air. The trachea extends downward and branches into two major bronchi, that travel further down the lower respiratory tract to form bronchioles and alveoli. At the top of the trachea is the cricoid cartilage that connects it to the larynx.

There are several C-shaped rings that are made of hard cartilage along the length of the trachea. These help to support the trachea so that it is virtually impossible for it to collapse under almost all circumstances besides major trauma. The rings are open in the back and there is a muscle, known as the *trachealis muscle*, that joins the ends. The rings are also joined to one another by bands of connective tissue that form the annular (circular) ligaments of the trachea. This is what the trachea and bronchi look like:

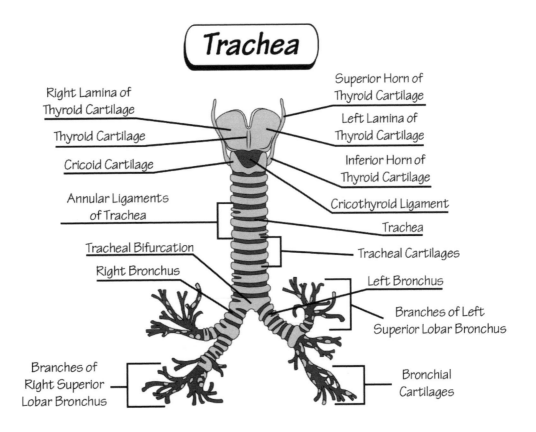

On the left there are two secondary bronchi, which are also called *lobar bronchi*. Each lobar bronchus goes to one of the two lobes of the left lung. On the right, there are three secondary bronchi—one for each of the three lobes of the right lung.

There are also tertiary bronchi. These are more like smaller city streets instead of major highways. There are eight tertiary bronchi on the left side of the lungs and ten tertiary bronchi on the right. These get smaller and smaller until they turn into the respiratory bronchioles.

Respiratory Bronchioles

The bronchioles are the smaller passageways through which air passes directly into the alveoli. They have no cartilage to support them or any glands within their submucosal layers. They form the last few parts of the conducting zone of the respiratory system. Bronchioles become smaller until they become *terminal bronchioles* that are also part of the conducting zone. These terminal bronchioles finally divide into even smaller *respiratory bronchioles* that are considered part of the respiratory zone. They terminate into the alveoli.

A *pulmonary lobule* is that part of the lung ventilated by just a single bronchiole. Bronchioles are 1 millimeter or less in diameter. Their walls are made from ciliated cuboidal epithelium and a layer of smooth muscle. The tinier bronchioles known as *terminal bronchioles* are less than 0.5 mm in diameter. Terminal bronchioles mark the very end of the conducting zone and the beginning of the respiratory zone.

The diameter of the bronchioles plays a crucial role in the flow of air. The bronchioles can change their diameter in order to reduce or increase airflow. An increase in diameter is known as *bronchodilatation* and is stimulated by the sympathetic nervous system.

Fun Factoid: What causes the bronchioles to spasm as you see in asthmatic conditions? The list is very long and includes allergen exposure, exposure to chemicals or irritants, exercise, stress, cold weather and infection. There are smooth muscle cells influenced by your autonomic nervous system that control whether or not bronchospasm *happens under these conditions. Bronchospasm is connected to inflammation, swelling and mucus production in the bronchial tree—all reasons why air you try to breathe in or out might not get where it needs to go.*

A decrease in bronchiolar diameter is known as *bronchoconstriction* and is stimulated by cold air, chemical irritants, parasympathetic nerves or histamine. A person with anaphylaxis has bronchoconstriction, with airways too narrow for adequate flow of air to the alveoli. When epinephrine or albuterol is given to these patients, these drugs stimulate the sympathetic nervous system receptors in order to dilate the bronchioles.

The epithelial lining in the bronchi starts as a simple columnar epithelium, with cilia that changes into simple ciliated cuboidal epithelium as the bronchioles become smaller. As the smaller bronchioles don't have cartilage, they use elastic fibers attached to the surrounding lung tissue that provide support for the airways. They are thin-walled tubes with no mucous glands present, and are surrounded by smooth muscle that will contract or relax, depending on the circumstances.

Terminal bronchioles represent the most distal aspect of the conducting zone. They are branches off the lesser bronchioles. Each of the terminal bronchioles divides to make respiratory bronchioles that contain a small cluster of alveoli. These terminal bronchioles contain ciliated cells but no goblet cells.

These tiny terminal bronchioles also contain *club cells* that have no cilia but are instead rounded protein-secreting cells in the epithelium. Their secretions are moist but not sticky. The compound made by their secretions allows for the bronchioles to get bigger during inspiration, and prevents the bronchioles from collapsing during expiration. Club cells, which are the major stem cells of the respiratory tract, also make enzymes that detoxify substances that might be dissolved in the respiratory fluid.

The respiratory bronchioles are the narrowest airways of the lungs, about 1/50th of an inch across. They are the result of multiple divisions of the bronchial tree. The bronchioles deliver air to the respiratory surfaces of the lungs. These tiny respiratory bronchioles stop at the alveoli, which are the thin-walled sacs where gas exchange occurs. The alveoli have ducts that connect to the respiratory bronchioles.

The Alveoli

A single air sac is called an *alveolus.* Together, they are called the *alveoli.* These are tiny grape-like structures where the exchange of oxygen really takes place. The membrane of the alveolus can be called an *alveolar membrane* or *respiratory membrane.* The walls are so thin here that gases like oxygen and carbon dioxide can freely move through them. As you can see by the picture, there is an extensive capillary layer around the alveoli in order to allow this gas exchange to occur:

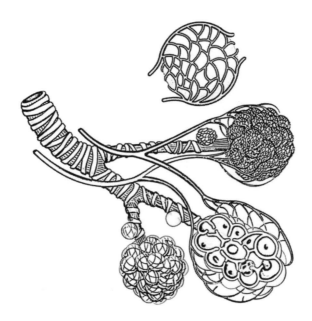

There are about 700 million of these alveoli in both lungs, with an extensive surface area of about 70 square meters. It takes this large a surface area to have enough space for gases to effectively get exchanged. The unique thing about alveoli is that they are elastic like a balloon. This means they can expand and collapse as air goes in and out of them. Like a balloon, they don't collapse completely when air leaves them.

The epithelium of the alveoli is not homogeneous. There is a thin cellular layer and a matrix made of elastin and collagen, but not any cells. This fluid matrix allows gases to dissolve in it so that diffusion across the respiratory membrane is easier. There are pores between adjacent alveoli called *pores of Kohn* that equalize the air pressure between the different air sacs.

There are three major cell types in the alveoli. Some of these are called *type I* or *squamous alveolar cells*. They are structural cells that don't have the capacity to regenerate, which makes them especially vulnerable to injury. *Type II pneumocytes* are also called *great alveolar cells*. These make surfactant, which is a thin, soapy fluid that keeps the alveoli from collapsing altogether. They can regenerate themselves. A third type of cell includes the macrophages, called *alveolar macrophages*, which provide immune defense against pathogens in the lungs.

Why the soapy surfactant in the alveoli? This surfactant is truly a type of detergent that reduces surface tension. It makes the alveoli a lot like soap bubbles. These eventually burst, but for a time the surface tension is decreased by the detergent in them, so that they stay bubbly and don't collapse right away.

Fun Factoid: What is surface tension and why is it important? Surface tension is the measure of the force necessary to expand a liquid surface at the air/liquid interface. Water has a high surface tension, so it takes a lot of force to try and expand it. The normal surface tension of water is 70 dyn/cm; but in the lungs, the alveoli have a maximum surface tension of 25 dyn/cm because of the presence of surfactant. As the alveoli collapse, the density of surfactant in them increases so that the surface tension is extremely low—nearly zero. That means it doesn't take nearly as much force to expand these alveoli as it otherwise would.

Respiration strictly means exchange of gases, while *ventilation* means the entire process of drawing air in and out of the lungs. Respiration is basically just the exchange of gases in the lungs, while ventilation is the delivery system and all of the things that go on backstage to support the respiratory process.

Ventilation is the Act of Breathing

Another term for ventilation is *breathing*. It's the process that must happen in order to get air into and out of the lungs. There are two phases to ventilation. They are inspiration, which is breathing in, and expiration, which is breathing out. As mentioned, it takes muscular effort in order to have inspiration, but little to no muscular effort in order to have expiration.

Before we get too far into this, there are some important facts you need to know about gases. The atmosphere around you is a mixture of gases. Surprisingly, most of it is *not* oxygen. The air we breathe at sea level contains about 78 percent nitrogen, 21 percent oxygen, slightly under 1 percent argon, and less than 0.04 percent carbon dioxide. Water vapor content varies according to how much humidity there is in the atmosphere. On a hot day with a dewpoint of 86 degrees, the water vapor percentage can be as high as 4 percent.

The atmospheric pressure depends on where you are on Earth. At sea level, the imaginary column of air pushing down on your head from the outer stratosphere is about 760 millimeters of mercury. But higher up on a mountain, for example, the air pressure is less than that. It's why we call this kind of air *thin air*. At the top of Mount Everest, the air you'd breathe is only 250 millimeters of mercury. This is why it's so hard to breathe up there.

Another thing you need to know is that all gases flow from high-pressure situations to low-pressure situations. You don't push air into your lungs when you breathe but instead expand your lungs, increasing the total volume in the thoracic cavity, which automatically decreases the pressure inside it. This allows the natural flow of air from high pressures near your nose to the low pressures in your alveoli.

This idea that larger volumes mean less pressure is called *Boyle's law of gases*. Gases are nothing more than floating atoms in a large space. If you increase the size of the space, these molecules can spread out and the pressure in the space decreases. By increasing the thoracic cavity size, airflow in the general direction toward the alveoli happens naturally. This picture helps to describe Boyle's law:

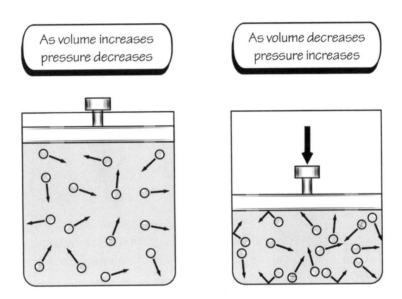

The lungs themselves remain stuck to the inside of the chest wall because of a continual negative pressure in the pleural space. This acts like a suction cup along with the pleural fluid to help form a seal between the lungs and the thoracic cavity, making this small space even more negative when it comes to pressure. If a hole were to develop in your chest wall or inside your outer lung tissue, your lungs would collapse on that side. When this happens, it's called a *pneumothorax*.

Fun Factoid: The most dangerous kind of pneumothorax is called a tension pneumothorax. Air gathers in the space between the thoracic wall and the lungs, so the lung will collapse. The problem with a tension pneumothorax is that air going into this extra space can't escape, so more and more keeps going into it. This will expand one side of the thorax greatly, pushing on the good side and blocking the flow of blood back to the heart. This is dangerous and often deadly. Doctors sometimes correct this by putting a needle into the chest wall in order to allow that excess gas to escape.

Inspiration and Expiration

Your diaphragm is a large muscle at the base of the thoracic cavity that contracts, going from a relaxed dome shape to a contracted flattened shape, which expands the lungs. The external intercostal muscles between each rib also contract. This action lifts up the sternum and ribs so that your chest expands from a front-to-back perspective. Both of these actions draw air into the lungs. This is what it looks like:

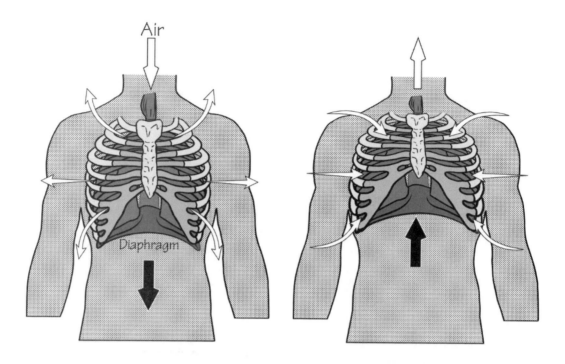

During passive expiration, these muscles simply relax and return to their resting positions. This will decrease the volume of the lungs so that air naturally flows out into the atmosphere again.

In forced breathing, such as when you are really exerting yourself, there are other muscles that extend the entire process and allow you to breathe in faster and take in more air in one breath. These muscles are called the *accessory muscles* of breathing. Many muscles are involved in this process, such as the scalenes, sternocleidomastoid muscles, pectoral muscles, latissimus dorsi, and serratus anterior muscle.

Resistance

When you suck up water from a straw, you have to use some suction power, right? The reason you have to do this is because there is resistance to the flow of water through the straw. The resistance is greater if you are sucking through a very narrow straw than if the straw has a larger diameter.

While gases are not as viscous as water, there is airway resistance exerted by the different tubes and structures through which the air passes when you suck air into your lungs and allow it to flow out again during expiration. The diameter of the different airways, such as that of the larger trachea and the smaller bronchioles, makes a big difference in how much airway resistance there is.

Another factor involved is whether the airflow is *laminar* or *turbulent*. Laminar flow is smooth. The molecules don't bump against the side walls or each other as much, and things flow more easily. In turbulent flow, the ride is rough. When a bronchus branches, there will be turbulent flow at the bifurcation and anywhere else that the lining of the bronchial tube is irregular. This is the basic difference between turbulent flow and laminar flow:

Turbulent

Laminar

According to *Ohm's law*, which applies to gases and fluids, the greater the resistance, the more pressure is necessary to allow the same airflow to reach the alveoli. According to *Poiseuille's law*, the diameter to the fourth power is inversely proportional to the resistance. This means that a tube that is four times as narrow as another tube will have 4^4 the resistance to flow of the same fluid or gas. This doesn't bode well for the very narrow bronchioles in the lungs. But because there are so many of these tiny bronchioles all together, this cuts down on the total amount of resistance you'll find in these parts of the lungs as a whole, even though the resistance in an individual bronchiole is high.

As we've discussed, the sympathetic and parasympathetic nervous system influence the narrowness of the small airways. The sympathetic nervous system will dilate the airways and reduce resistance, improving the flow of air to the alveoli. If the parasympathetic nervous system is active instead, there will be the opposite reaction and you will have bronchoconstriction that decreases airflow to the alveoli. This happens because of changes in the smooth muscle that controls the diameter of the airways.

Inspiration causes positive pressure in the alveoli and small airways, so resistance to airflow will be less. In expiration, this isn't the case and airflow resistance is greater. This is often why asthmatics experience wheezing when breathing out but not so much when breathing in. The cartilage in the larger airways helps to prevent total collapse. The problem is much worse during forced expiration of air out of the lungs.

If the inside of the lungs was just dry, the airways would be elastic and stretchy. If the lining of the airways was very watery, the high surface tension of water would increase resistance, because it tends to draw the airways into a more narrowed state.

Respiratory surfactant, on the other hand, does not have this high level of surface tension, so the alveoli in particular will expand easier. This allows air to get inside them to a greater degree. Surfactant gets spread out on larger alveoli, making it less effective during expansion. Surfactant is less spread out in alveoli that aren't expanded, so the surface tension will be very low, allowing them to expand more easily the next time you take a deep breath.

Understanding Lung Volumes

You might think that all the air in the lungs gets magically transformed in the alveoli, where all the necessary oxygen is sucked into the body and all the carbon dioxide is gathered for waste removal by the lungs. This isn't what happens at all. Just as in any piping system, there is some air that stays in the pipes and never makes it to the alveoli. This gives rise to a natural dead space, which is the amount you breathe in that never gets exchanged. This dead space can increase in certain diseases, such as *emphysema*, which increases the work of breathing and leads to lower oxygen levels.

These are the terms that respiratory physiologists use to describe ventilation:

- **Tidal volume:** The amount breathed in during quiet breathing in total
- **Inspiratory reserve volume:** The amount you can maximally breathe in past the tidal volume
- **Expiratory reserve volume:** The amount you can force out beyond the tidal volume during maximal expiration
- **Residual volume:** The amount left in the lungs even when you breathe out completely, which is about 1.5 liters
- **Forced vital capacity:** The total amount you can inhale after maximally inspiring air, which is roughly 4.5 liters
- **Inspiratory capacity:** The tidal volume plus the inspiratory reserve volume
- **Functional residual capacity:** The amount remaining after you quietly breathe out, which is approximately 3 liters
- **Total lung capacity:** The total volume you can breathe in maximally, which is about 6 liters

The *anatomical dead space* is the amount of air that never reaches the alveoli at all because it remains in the tubes. The *alveolar dead space* is the amount of air that gets to the alveoli but cannot participate in gas exchange.

This is a rough idea of these different lung volumes:

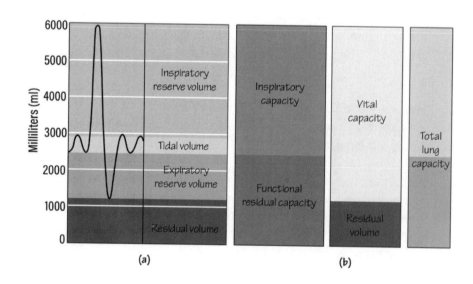

What Controls Ventilation?

Remember that ventilation is all about mechanics. Air flows in and air flows out. This process must work fairly well if air is to even get into the alveoli in the first place. There are factors that play into why you breathe fast sometimes and slow at other times. They also control your ability to take a deep breath or a shallow breath.

One way to have control over ventilations is through the use of *chemoreceptors*. These are receptors in the *carotid* body and aortic body that detect the oxygen and carbon dioxide levels in the blood. If they detect too little oxygen or too much carbon dioxide, you will be triggered to breathe faster and deeper in order to improve these numbers. The receptors have a direct connection to the brainstem respiratory centers. These send out their own signals to direct how fast and how deeply you breathe. Your cardiac output also increases to get more oxygen to the tissues.

There are also central chemoreceptors in the *medulla oblongata* of the brainstem. These are actually more sensitive to carbon dioxide and oxygen levels in the brain, and will directly affect the respiratory centers that control how much you need to breathe. If you don't breathe enough, carbon dioxide builds up and your blood becomes more acidic.

Of course, all of this happens involuntarily, and is how the subconscious parts of the brain keep you breathing even while you sleep. Separate groups of neurons in the medulla and *pons* of the brainstem send out breathing signals to allow for a rhythmic breathing pattern. Some people have what's called *central sleep apnea* where these respiratory centers fail for a period of time, and the person stops breathing in their sleep.

Fun Factoid: The sad fact is that a common cause of ventilatory failure and death in this country is opioid overdoses or overuse. High opioid levels turn off the brain's respiratory centers, and your brain stops sending signals that tell you to breathe.

Of course, there is also voluntary control over ventilation, because you can breathe faster or slower when you want to. There are signals from the emotional centers of the brain and the *hypothalamus* that dictate your respiratory rate in a more voluntary way. There are also motor or movement fibers in your brain that get triggered to allow you to voluntarily breathe at a set rate.

Finally, there are *stretch receptors* in your lung tissue that tell you when you have breathed deeply enough. These are also in the bronchial tree so that you don't overstretch your lungs during the act of breathing.

Respiration and Gas Exchange

Respiration is the process of gas exchange across a membrane. Because this process happens both in the lungs and at the tissue level, we sometimes refer to alveolar respiration as *pulmonary respiration* in order to distinguish it from tissue-based *peripheral respiration*.

Remember that gases exist in a mixture in the air that reaches the lungs and enters the alveoli. There is a total pressure of all the molecules together, and partial pressures exerted by the different gaseous components separately. According to *Dalton's law of partial pressures*, you can add up the different partial pressures of a mixture to get the total pressure.

Gases in liquids are not really gaseous but are dissolved in the liquid somehow. Some, like nitrogen, are present to a great degree in a gas, but will not diffuse into a liquid medium because they don't dissolve very well in liquids like blood. This means that nitrogen isn't found to a great degree in the bloodstream as a dissolved substance.

Fun Factoid: Have you ever heard of the bends? This disorder affects divers who ascend to the surface too quickly after a deep dive. When you descend to deep levels under water, the pressure around you is greater, so nitrogen dissolves in your blood. Once you ascend, the nitrogen is not as dissolvable and it forms bubbles within the blood. Bubbles of nitrogen travel to parts of your body, preventing your circulation from reaching these areas. It is called the bends because the joints are primarily involved in the resulting symptoms.

Carbon dioxide dissolves much better, so it can flow from a gas to a liquid more easily, as long as the concentration in the blood exceeds the partial pressure of carbon dioxide in the alveoli. Oxygen is found in a higher concentration in the alveoli and the trend is to flow into the blood from the alveoli. This is made more efficient by the fact that hemoglobin grabs up these oxygen molecules, which effectively mask them from the equation so that even more oxygen can get dissolved into the bloodstream as a free gas.

One of the three things that most affect the diffusion of gases in any part of the body includes the concentration gradient. If the gradient, or difference in concentration across the membrane, is greater, the rate of diffusion will be faster. In addition, if the surface area where diffusion should take place is great (as is true of the healthy alveoli), the diffusion will be faster as well. If the pathway across the membrane is thick enough, this will reduce the rate of diffusion. Tinier gas molecules diffuse more quickly than large gases.

There is also *Henry's law*, which applies to the behavior of gases in liquid-like plasma. If the partial pressure of the gas is high, the amount of gas that will dissolve in the liquid will be high as well. *Fick's law* is more complex and says that there are a lot of different factors that affect diffusion of a gas through a liquid. These include the gas's solubility, the cross-sectional area of this fluid, the partial pressure differences across a membrane, the distance it must take to diffuse in the first place, the size of the gas particle, and the fluid temperature.

Carbon dioxide diffuses twenty times faster across the respiratory membrane than oxygen, even though oxygen gas is smaller in size per particle. This difference gets compensated for because there is a much greater concentration gradient for oxygen across the respiratory membrane. The downside, though, is that in lung diseases, it's the oxygen exchange that suffers more than the carbon dioxide exchange. Hemoglobin helps because it sucks up the oxygen as rapidly as possible, in order to drive diffusion toward being taken up by the alveoli.

This is what it looks like in the alveoli when respiratory gas exchange occurs:

Function of the Alveolus in the Lungs

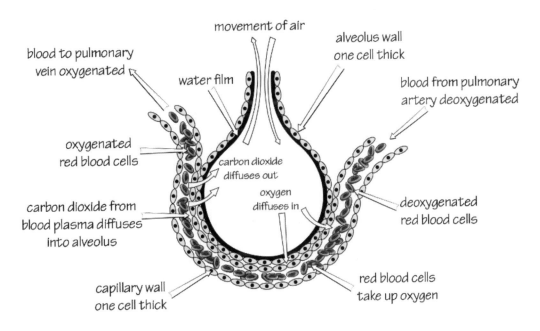

Because blood flow increases so fast during exercise, the rate of diffusion increases to about half a second after a blood cell gets into position along the respiratory membrane. What this means is that exercise is not limited by the amount of gas exchange occurring across this membrane. You can exercise and still have the ability to take on oxygen in your lungs.

Carbon dioxide has a high partial pressure in the capillaries, after getting dumped into the circulation as a metabolic waste product. This affects the diffusion capacity because there isn't much carbon dioxide in the alveolar gases. Once this diffusion happens, you breathe out the carbon dioxide and breathe in new air that's low in carbon dioxide. This starts the whole process all over again. Most carbon dioxide is directly dissolved in blood plasma. Some of it gets attached to proteins, while some undergoes a chemical change to become dissolved carbonic acid. This is why carbon dioxide is considered an acidic substance in the blood.

Ventilation and Perfusion Coordination

As we've discussed, blood flow must happen directly in the same area as the alveoli full of oxygen, or the normal exchange of gases can't occur. This process isn't ever perfect, even in normal persons. There is a ventilation rate that involves the volume of gases inhaled and exhaled during a certain time period. This is the tidal volume multiplied by the respiratory rate per minute. Lung perfusion is the amount of blood getting to the pulmonary capillaries in the same time period.

In a perfect system, the ventilation/perfusion ratio would be 1. This is also called the *V/Q ratio*. This doesn't happen in reality. When you stand up, the ratio is high in the upper lungs—as high as 3.3—while in the base, this same ratio is 0.6. There is more ventilation in the top of the lungs and more perfusion in the base of the lungs. Gravity has a lot to do with this. Averaging these ratios out gives an overall ratio of about 0.8.

In some cases, there is a larger V/Q mismatch, which can lead to *hypoxia*, or low oxygen levels in the bloodstream. It can happen if the alveoli are damaged or too thick to diffuse oxygen. It can also occur in pulmonary emboli if blood flow is blocked to parts of the lungs. The body will compensate by trying to send blood only to the better-ventilated lung areas, but this also isn't a perfect system.

Hemoglobin

Hemoglobin is a four-peptide-chain globular protein complex that can carry up to four oxygen molecules at a time. It is absolutely necessary for survival, because oxygen by itself isn't as dissolvable in plasma as it needs to be in order to get oxygen to the tissues. Hemoglobin is a dynamic molecule that is about 99 percent saturated in arterial blood at any point in time. Surprisingly, it also carries carbon dioxide and is called *carboxyhemoglobin* whenever it does this.

The whole process is not as simple as it looks like it should be. Hemoglobin participates in what's called *cooperative binding*. That means that it's hardest to get the first oxygen molecule to attach to hemoglobin, but when it happens, the rest of the oxygen molecules become attached much more easily. Empty hemoglobin is called *tense-hemoglobin*, or *T-hemoglobin*, while hemoglobin with just one oxygen attached is called *R-hemoglobin*, or *relaxed hemoglobin*.

Relaxed hemoglobin draws in oxygen better than tense hemoglobin. This is an advantage in the alveoli, where oxygen tension is high. Oxygen is taken up very efficiently in this situation. In the tissues where oxygen tension is low, however, the push is instead to draw off as much oxygen as possible when it is most needed.

Hemoglobin's affinity for oxygen isn't the same in all circumstances. There are certain environmental factors that affect just how much oxygen will attach to hemoglobin, like the pH level. When the pH is low in the tissues, it means a lot of metabolism has been going on and the need for oxygen is greater there. This drives less oxygen to attach to hemoglobin. High temperatures generated by moving muscles also lower the affinity of oxygen for hemoglobin, so that muscle cells get more oxygen when active. This means you will offload oxygen better in these metabolically active tissues.

A molecule called *2,3-DPG*, or *2,3-diphosphoglycerate*, is found in RBCs and will also lower the ability of oxygen to attach to hemoglobin. This chemical increases in RBCs as a person acclimates to higher altitudes, in order to make hemoglobin-oxygen transport more efficient.

Carbon Dioxide in the Blood Plasma

The *Haldane effect* influences the ability of carbon dioxide to attach to hemoglobin. If the oxygen level is too low in the tissues, for example, the hemoglobin will be much more likely to attach to carbon dioxide instead. This is how hemoglobin picks up carbon dioxide at the tissue level. More than 60 percent of carbon dioxide gets transported as bicarbonate ions, or $H_2CO_3^-$. An enzyme in RBCs called *carbonic anhydrase* does this. By sending bicarbonate ions into the blood stream, it can help buffer the blood so there aren't so many extremes in pH.

About 10 percent of the CO2 in the plasma is directly dissolved as a gas. This is very different from oxygen, which really needs hemoglobin for this process. Carbon dioxide is considered 23 times more soluble in plasma than oxygen. At the tissue level, there's a lot of carbon dioxide in the cells. This gets diffused out into the capillaries and is either taken up by hemoglobin, dissolved in solution, or made into bicarbonate to be carried away.

The Lungs and Acid-Base Issues

Blood plasma doesn't tolerate extremes of pH very well, mostly because proteins and enzymes are super sensitive. The lungs are capable of very quickly helping this process of keeping the pH stable simply by changing breathing patterns. Of course, there are buffering systems that chemically affect the rate of change of pH in the blood plasma, and the kidneys do their own job. The main difference between the lungs and the kidneys as buffering systems is that the lungs are fast and efficient at making these pH changes as well.

The process of regulating pH in the blood plasma is called *compensation*. If the chemoreceptors in the periphery or in the brain detect a pH or carbon dioxide change that is not within normal limits, it causes the lungs to either hold onto carbon dioxide or blow off carbon dioxide. How does it work?

You hold onto carbon dioxide if you don't breathe as fast or as deeply. This means less carbon dioxide exchange in the lungs and more carbon dioxide in the plasma. Carbon dioxide (by virtue of its relationship to carbonic acid) is considered acidic and will lower the pH. In the same way, if you breathe fast and deeply, carbon dioxide is blown off the lungs and into the atmosphere. This will have the effect of raising the pH of the blood plasma.

In disorders of *respiratory alkalosis*, you breathe too much or too fast. You essentially hyperventilate. The end result is a rapid rise in blood pH as carbon dioxide is blown off. If, on the other hand, you take a drug like an opiate that suppresses ventilation, you don't breathe enough and carbon dioxide builds up. This leads to *respiratory acidosis*. The combination of low oxygen and respiratory acidosis is what makes opioid overdoses so dangerous and deadly.

Fun Factoid: People who have chronic obstructive pulmonary disease often have chronically low oxygen levels. Their brain essentially switches to detect oxygen concentration as its major drive to breathe. The problem happens when you give people with COPD a lot of oxygen in an attempt to bring up their oxygen levels. Their brains see that the oxygen level is good and shut off the respiratory centers. In a sense, too much oxygen can cause them to quit breathing and to collapse into respiratory failure.

To Sum Things Up

The respiratory system is absolutely essential to life. Without its ability to take in oxygen on a regular and consistent basis, the aerobic metabolism your cells operate on could not happen. The whole process involves ventilation so that air gets into and out of the alveoli of the lungs. It also involves pulmonary respiration, which is the actual act of gas exchange in the alveoli.

The entire thing is like an integrated delivery system that coordinates with the circulatory system. There are conducting airways that act like large thoroughfares for the passage of air in and out of the lungs. Their job is to deliver the oxygen, which is precious merchandise for your cells. These large airways become smaller and smaller before ending in the alveoli, where the merchandise is delivered to capillaries that surround each alveolus.

The whole process must be well-coordinated. Oxygen must go to as many of the alveoli as possible, and the alveoli must be healthy enough to allow oxygen and carbon dioxide to be exchanged. The circulation, or perfusion, to the lungs must coordinate with the respiratory tract, so that the areas where oxygen can be found are the same as where the capillaries are located. If these two systems aren't well integrated, oxygen and carbon dioxide won't be effectively exchanged by this system.

Chapter Ten: Eating and Your Gastrointestinal System

The main goal of your gastrointestinal, or GI, system is to take in nutrients from your diet, digest them (which is the same thing as breaking them down into small molecules), and absorb the nutrients for use by the body's cells. An added job equally important is to eliminate the wastes that couldn't be digested or absorbed, and to regulate water excretion to a minor degree. The whole GI system involves numerous organs of the mouth, thorax and abdomen. The main part of this system is the *alimentary canal*, although the liver, gallbladder and pancreas are also important GI organs.

Anatomy of the Alimentary Canal

The alimentary canal is also called the *digestive tract*, the pathway by which food enters the body, gets processed and absorbed, and gets eliminated if it's not directly used. The major parts of the alimentary canal include the mouth, pharynx, esophagus, stomach, small intestine, large intestine and anus. This image shows your digestive system as a whole, with your alimentary tract as the major feature of this system:

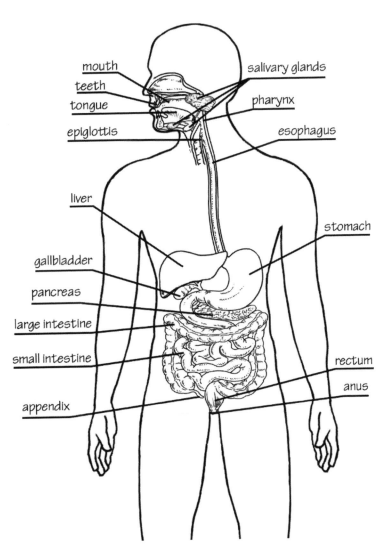

The Mouth

The mouth is also called the *oral cavity* or *buccal cavity*. In the gastrointestinal tract, food enters the mouth and is chewed and ground down by the teeth. The tongue moves the food around the mouth and helps to coat the food with saliva, so that it can pass through the esophagus as a moist *bolus*. The tongue also helps to move food back so it can be swallowed.

The boundaries that help define exactly where the mouth is located include the lips at the mouth opening, the hard and soft palates at the roof of the mouth, the glottis in the back, and the cheeks, or *buccal sides*, on either side. There are two parts to the mouth: the *vestibule*, which is the area between the teeth and the cheeks, and the oral cavity itself. The oral cavity is mostly filled up by the tongue, which is a large muscle connected to the floor of the mouth by means of the *frenulum linguae*.

Digestion is of two types, mechanical and chemical. *Mechanical digestion* is the physical breakdown of food. In the mouth, the main structures that do this are the teeth, which grind and chew ingested food into smaller pieces, and the tongue, which mixes food and also carries sensory receptors used for taste.

The mouth also participates in *chemical digestion* to a lesser degree. There are glands in the mouth that make an enzyme called *salivary amylase*. This is a minor enzyme in the overall digestive process, but it does start the process of breaking down carbohydrates. There is also an enzyme made in the mouth called *lingual lipase*, which starts breaking down the fats you take in.

The mouth is completely lined by stratified squamous epithelium that contain many different glands. These, along with the salivary glands, moisten the food so it can begin the process of digestion. There are specialized surfaces that form the covering of the gums that support the teeth and the tongue surface. The surface of the tongue has a rough texture because of the taste buds that are spread out across its surface.

Pharynx

We've talked about the pharynx with respect to the respiratory tract, but it's just as important for the digestive system. This is basically a cone-shaped passageway that leads from the nasal and oral cavities inside the head to the larynx and esophagus. When it comes to digestion, the pharynx is not passive. It has muscles that push food from the mouth down into the esophagus in small boluses in the act of coordinated swallowing. The epiglottis is at the base of the pharynx. The act of swallowing closes the epiglottis so that food passes down the esophagus and doesn't enter the trachea. This is the epiglottis, which is located above the larynx:

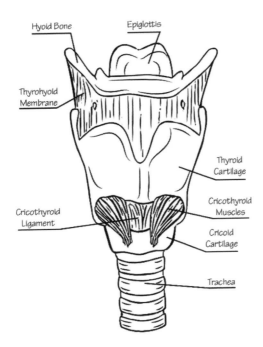

As you know already, there are three main divisions to the pharynx. There is the upper portion, the nasal pharynx, located at the back of the nasal cavity. The nasal cavity then connects to the oropharynx, which begins at the back part of the mouth cavity and ends in the epiglottis. There are two recesses in this part of the pharynx for the palatine tonsils, which are areas of lymphatic tissue that are often prone to infection.

The isthmus between the nasopharynx and oropharynx is important because it allows humans to breathe through either their mouth or their nose, and also allows for food to pass through the nose by means of a nasogastric tube in cases where a person cannot eat for themselves. The third and lowest region of the pharynx is the laryngopharynx, which begins at the epiglottis and leads down to the level of the esophagus. It regulates the passage of air into the bronchial tree and food through the digestive tract.

The Esophagus

The esophagus is a relatively straight tube that passes food from the pharynx to the stomach. The esophagus has muscles that allow for expansion and contraction of its walls in order to facilitate the passage of food. In that sense, it's not just a passive tube but has *peristaltic* activity that pushes food through its lumen.

Peristalsis is a process that happens throughout most of the alimentary tract. Think of it as what you do when you squeeze a tube of toothpaste. You squeeze at one end and the toothpaste pops out the other end. With peristalsis the process is similar, but happens in a rhythmic fashion that continually pushes food down from the top of the esophagus to the rectum.

The esophagus lies behind the heart and trachea and is directly in front of the spinal column. It goes through the diaphragm, where it connects to the stomach. Both ends of the esophagus can close off. There is an *upper esophageal sphincter* at the top of the esophagus and a *lower esophageal sphincter* at the bottom of the esophagus. The upper sphincter is made from a circular muscle tissue and is generally closed. When food enters the pharynx, this sphincter relaxes so that food can enter the esophagus. After a bolus passes, the sphincter closes to prevent food from traveling back into the pharynx.

The first third of the esophagus has mainly skeletal muscle, which allows it to participate in the voluntary act of swallowing. The middle third of the esophagus is a mixture of skeletal and smooth muscle fibers, while the lower third is almost all smooth muscle. Smooth muscle is involuntary but still undergoes peristalsis. The food finally reaches the lower esophageal sphincter, which then opens up to allow for the passage of food into the stomach. Afterwards, it closes to prevent the gastric juices from the stomach from backing up and injuring the esophageal mucosa.

The Stomach

The stomach is the only significant receptacle for a large amount of food in the alimentary canal. It is a greatly widened area of the digestive tract, located between the esophagus and the small intestine. In the upper abdomen, you can find it in the anterior or front of the abdominal cavity, with the pancreas nestled behind it and the left kidney and spleen located nearby. This is what it looks like in situ:

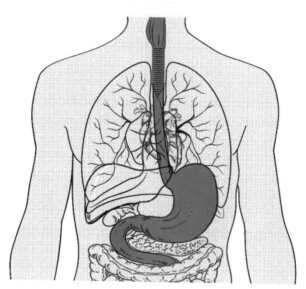

The stomach is amazingly distensible. When you don't have any food in it, it's about 6 by 12 inches in width and length. In its relaxed state, the stomach will hold only about 2.5 ounces; but when you eat a lot, your stomach can safely relax to hold more than a quart of food at once. Because the small intestine cannot digest nor absorb food that quickly, the stomach holds its contents so that it can gradually release it into the duodenum when it's ready to handle more.

This organ's main jobs are to store food prior to absorption in the small intestine, and to secrete *hydrochloric acid*, which breaks down proteins into amino acids so they can better be absorbed later on. In addition to this chemical digestion process, it also mechanically breaks down food.

Parts of the stomach include the *cardia*, the *fundus*, the body of the stomach, the *pyloric antrum*, and finally the *pyloric canal*. The pyloric canal ends in the *pylorus* that sends partially digested food into the esophagus. This partially digested food is called *chyme*. This is what the different parts look like:

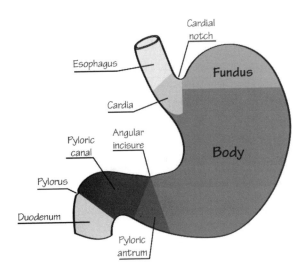

The Small Intestine

The small intestine is a narrow, folded or coiled structure that extends from the stomach to the large intestines. It is the part of the digestive tract where most of the digestion and absorption of food takes place. The "small" part of this structure refers to its caliber but not its length. It is actually the longest part of the alimentary canal, with a length of about 22-25 feet. In addition, the interior is highly convoluted, with *papillae* and *microvilli* that increase the absorptive surface of the mucosa. These structures extend the surface area to about 2,700 square feet, or roughly the size of a regulation tennis court.

The small intestine has three separate segments—the duodenum, the jejunum, and the ileum. It can be found coiled up in the middle and lower aspects of the abdomen. It is both covered and somewhat suspended in place by the *mesentery*, a fatty protective membrane found in nearly all of the abdomen.

The duodenum is a short segment of small intestine in the shape of a C. It carries out more digestion than it does absorption of nutrients, mostly because it contains openings from the gallbladder, liver and pancreas, which all participate in some way with digestion. While the stomach is a highly acidic environment, the duodenum has an alkaline environment, largely because bicarbonate is secreted into it as soon as food enters. Pancreatic enzymes and *bile* from the liver and gallbladder add to the food in order to help digest it.

The jejunum has the greatest absorptive ability of the entire small intestine. It participates in two kinds of peristalsis. Some of it naturally propels food from beginning to end, while other peristaltic movements actually churn food back and forth, keeping it in one spot so that absorption can happen maximally before the chyme is allowed to move forward.

The ileum is the terminal segment of the small bowel, or small intestine. It will absorb whatever nutrients haven't been absorbed by the jejunum, but is mainly responsible for absorbing vitamin B12 and bile salts, which get sent back to the liver after doing their own digestive functions in breaking down fats.

The Large Intestine

The large intestine makes up the distal aspect of the intestinal tract. It is made of four separate regions: the *cecum*, the colon, the rectum and the anus. The colon is divided itself into the ascending, transverse, descending, and sigmoid segments. The entirety of the large intestine is sometimes just called the *colon* or *large bowel*. Its function is to store bacteria that use undigested food material to make nutrients for the body, and to absorb water from the chyme so that it becomes more solid in the form of stool. This is what the large intestine looks like:

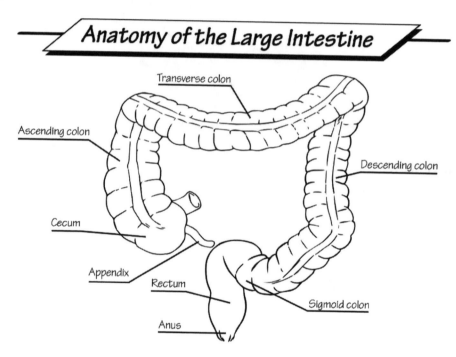

The large intestine is shorter and wider than the small intestine and is about 5 feet in length. It is only called "large" because of its larger caliber compared to the small intestine. It has a smooth inner lining compared to the highly convoluted internal texture of the small intestine.

Bacteria in the colon are also called your *gut microbiome*, which is crucial to your overall health. They make B vitamins and other nutrients that are then used by the body. There is a symbiotic relationship between these bacteria and the rest of your body. You have nutrients the bacteria need (which come from food you don't digest yourself), and bacteria make nutrients you are not able to make without them.

The colon absorbs a great deal of water every day. In order for the chyme coming from the stomach to absorb in the small bowel, it needs to be in a highly watery environment. After absorption, though, this water is not necessary, so it gets reabsorbed in the colon so that the chyme becomes stool.

The anus is the terminal end of the anal canal, which is the part of the digestive tract that passes any aspect of the food you have eaten that was not digested. Its main function is to have a sphincter that makes sure you can control your bowels and *defecate*, or pass stool, only when it is convenient to do so. A major disorder related to the anus is hemorrhoids, which are just anal veins that become dilated when you strain a lot while passing stool, are pregnant, or must stand or sit a lot without moving around.

Fun Factoid: What are hemorrhoids, anyway? These are nothing more than varicose veins around your anus. Veins become varicose because they are under too much pressure. If you have constipation and strain a lot to have a bowel movement, or if you are pregnant and the baby is pushing onto the rectal area, it can cause the veins to become varicose. They might bleed because they are fragile, or become painful if a blood clot develops in one of them. Sometimes a doctor has to lance these hemorrhoids in order to remove the blood clot, which makes the hemorrhoid look like a big purple grape.

The Peritoneum

The peritoneal cavity is the space where the different abdominal organs are located. Unlike the pleural and pericardial cavities, the peritoneal cavity is not a potential space but actually has fluid in it to a small degree, in addition to organs. It is bound by the peritoneal membrane. Like similar serous membranes, it has two layers that actually fold back upon themselves. The inner visceral layer covers each of the organs it comes in contact with, while the outer parietal layer lines the inside of the abdominal cavity.

Think of this peritoneal layer as a kind of blanket that covers the abdominal organs rather than a sleeping bag the organs lie within. It leaves some of them only partially covered on one surface or not at all. *Intraperitoneal structures* are completely wrapped up in the peritoneum, while *retroperitoneal structures* are behind this blanketing layer and aren't wrapped in it at all.

The most unique structures linked to the peritoneum are the *mesenteries*. These are folded areas that carry the nerves and blood vessels to and from the different organs. There are several mesenteries named according to the organs where they attach. The *transverse mesocolon*, for example, is the supporting mesentery of the transverse colon.

The peritoneal cavity is one space divided into two communicating compartments, the greater sac and the lesser sac. The greater sac is bigger and starts at the level of the diaphragm, extending down to the pelvic cavity. The transverse mesocolon divides it into an upper supracolic compartment and a lower infracolic compartment.

The lesser sac is also called the *omental bursa*. It is behind the stomach and in front of both the duodenum and pancreas. It is more movable in general, so that the stomach can shift about in the abdomen. The omental foramen is the opening between the greater and lesser omental sacs. This figure shows what these omenta look like:

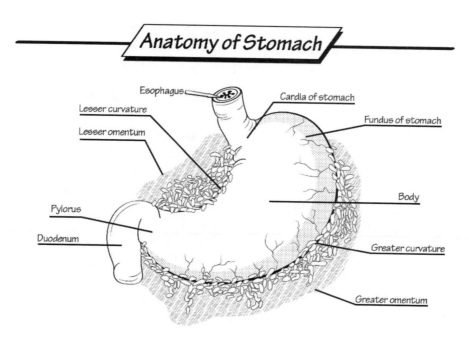

Anatomy of Stomach

Esophagus

Cardia of stomach

Lesser curvature

Fundus of stomach

Lesser omentum

Body

Pylorus

Duodenum

Greater curvature

Greater omentum

The greater omentum is the peritoneal fold that attaches the stomach to the transverse colon, while the lesser omentum attaches both the stomach and the duodenum to the liver. The greater omentum acts like an apron, covering most of the abdominal contents. It is fatty and membranous, carrying a great many nerves and blood vessels within it. Another of its jobs is to keep the visceral and parietal peritoneal surfaces from sticking to one another. It is a mobile organ and, if there is an inflamed part of the abdomen, it moves to cover it and protect the healthier parts of the peritoneal space.

The lesser omentum has two major ligaments within it: the *hepatogastric* ligament and the *hepatoduodenal* ligament. These serve to attach the upper abdominal organs to one another so that they stay in space. In addition, the lesser omentum has blood vessels and nerves that travel to the different upper abdominal organs.

The Walls of GI Tract Organs

The gastrointestinal wall is a multi-layered structure surrounding all organs of the alimentary tract. Like the walls of the different types of blood vessels, the alimentary tract wall has similarities and differences, depending on where you look. The different layers from the inner lumen outward are the mucosa, the submucosa, the muscular layer and the *serosa*. The outer serosal layer is called one of two different things. It is called the *serosa* if the GI segment is intraperitoneal, but *adventitia* if the segment is retroperitoneal.

The epithelium is the tissue that lines the inside of the GI tract, and is part of the mucosa. It is called *glandular epithelium* because it has a lot of goblet cells in them that make mucus. This mucus is necessary for digestion because it lubricates the food, allowing it to pass more freely through the gastrointestinal tract while also protecting the inner lining from damage by digestive enzymes.

In the small intestine, the mucosa looks different from the rest of the GI tract. It has many folds called *villi* that act to increase the surface area of this part of the intestinal tract. The villi contain something called a *lacteal*, which helps in the absorption of lipids and tissue fluids. There are microvilli on the surface of the large villi that massively increase the absorptive surface area over which absorption is able to take place. This is what the surface of the small intestinal lining looks like:

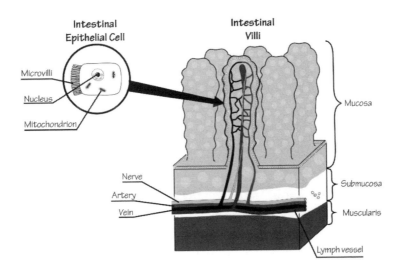

In the large intestine, there are many intestinal glands present in the tissue underlying the mucosa. There are no villi in the large intestine, but there are thousands of glands that secrete mucus. Underlying the mucosa is the *lamina propria*, which contains immune cells, nerves, blood vessels and myofibroblasts. Below that is the *muscularis mucosa*, a layer of smooth muscle that helps pass food through the gut by undergoing peristaltic contractions.

The submucosa in all parts of the GI tract contains the major nerve supply to the alimentary canal, often forming a ring around the canal called the *submucous plexus* or *Meissner's plexus*. It also contains the blood vessels and collagen-containing elastic fibers that stretch to make room for the digestive contents and maintain the shape of the intestinal wall.

Outside the submucosa is the muscular layer, consisting of circular and longitudinal smooth muscle layers that help move digested substances along the gastrointestinal tract. In between the two layers of muscle is the *myenteric plexus*, or *Auerbach's plexus*. This myenteric plexus is where the motor neurons wrap around the alimentary canal in order to cause peristaltic contractions.

The outermost layer of the alimentary canal is the serosa or adventitia. It is made of loose connective tissue and is coated in mucus, so that friction cannot damage the intestine should it rub against any other organ in the abdomen. Holding the gastrointestinal tract in place and connected to this layer are the mesenteries, which hold up the intestine and prevent disturbance of the intestine when you are physically active.

The Liver

The liver is an important organ of the gastrointestinal system. It is located in the right upper quadrant of the abdomen, just below the diaphragm. The liver has a wide range of functions that include protein synthesis, the detoxification of metabolites, the production of bile, and the processing of nutrients absorbed by the gastrointestinal tract.

The liver is considered a type of digestive gland and plays a huge role in many aspects of metabolism. Some of its many functions are to take up glucose and store it for future use in the form of glycogen, break down old red blood cells, synthesize most of the plasma proteins and clotting factors, produce hormones, and detoxify harmful substances in the bloodstream.

With regard to the digestive system itself, the liver is what makes bile, an alkaline substance that is necessary for the digestion of lipids. It sends the bile into the gallbladder, which is a pouch just beneath the liver. The gallbladder stores it and contracts whenever a person eats a fatty meal, in order to better digest the fats.

The liver is a wedge-shaped, reddish-brown organ that consists of four separate lobes. A normal human liver weighs about 3.2 to 3.7 pounds, making it the largest and heaviest gland in the human body. It is connected to two big blood vessels, the portal vein and the hepatic artery. The hepatic artery carries highly oxygenated blood from the aorta. This is also the blood vessel that supplies blood to the liver cells.

The portal vein is the confluence of many smaller veins that arise in the walls of the alimentary canal, particularly in the area of the small intestine. Its job is to carry blood high in digested nutrients from the GI tract as well as blood from the pancreas and spleen. The portal vein breaks into smaller and smaller segments until they become tiny capillaries in the liver known as *liver sinusoids*.

Each liver sinusoid leads to an individual *hepatic lobule*, the functional unit of the liver where all the action takes place. A single lobule is made from millions of liver cells, or *hepatocytes*, which are the basic metabolic cells of the liver. Lobules have a unique arrangement with blood flowing into each one of them continuously. This is what a lobule looks like:

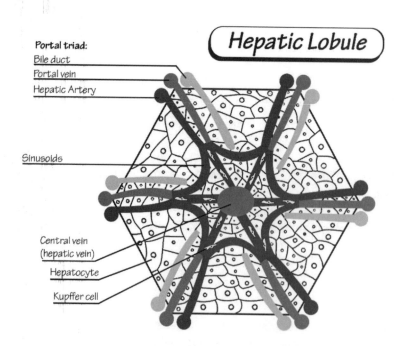

You can see the unique arrangement of the hepatic lobule. It allows for blood to be brought in and nutrients or toxins to be taken up by the hepatocytes in the lobule. There is the efficient inflow of blood to supply oxygen to these cells along with the input from the portal venous system. There is also output to the biliary tract, so that the whole thing is like a mini-factory that can do everything your body needs in a compact and highly efficient way.

The liver is divided into two segments, a right and left lobe. The two lobes look obviously different from one another and are separated by the *falciform ligament*. On the underside, though, you can see that there are actually four separate lobes, including the left lobe, the right lobe, the *quadrate lobe* and the *caudate lobe*. The quadrate lobe and caudate lobe cannot be seen except from underneath the liver itself. This is what the liver looks like anatomically:

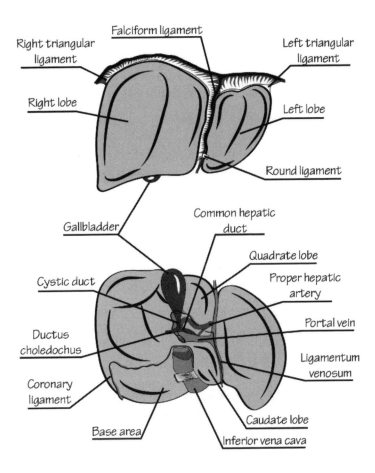

The visceral surface of the liver is on the underside of the organ. It is concave and has an uneven surface. Like the rest of the liver, it is covered in peritoneum except where it attaches to the *porta hepatis* and the gallbladder. The porta hepatis is a fissure beneath the liver that contains almost all of the major arteries, veins, nerves and hepatic ducts entering or leaving the liver itself. The inferior surface of the left lobe of the liver is what's known as the *gastric impression*. This is the impression that indents at the place where the stomach is located.

The Gallbladder

The gallbladder is also called a *cholecyst* or *biliary vesicle*. It is a small organ of the digestive system where bile is kept in storage and further concentrated before being released into the small intestine to aid in lipid digestion. While the gallbladder is important to digestion, you can live without it. When it is removed for any reason, the liver simply takes over and secretes bile that just isn't as concentrated as bile coming from the gallbladder.

The gallbladder is located just beneath the right lobe of the liver. It needs to be located there so that it's close enough to the liver to take up bile. In your body, this hollow organ is about 3 inches in length and 1.6 inches in diameter when it's full. It holds about 100 ml of bile. It looks a lot like a deflated balloon, with the open end marking the beginning of the *cystic duct* and the *biliary tree*. It is divided into three parts: the fundus, the body and the neck.

The fundus faces the front part of the body and represents the rounded end of the gland. The body connects to the liver and lies in the *gallbladder fossa*, which is an indentation at the bottom portion of the liver. The neck tapers off and is continuous with the cystic duct, through which the bile passes into the biliary tree. The cystic duct connects with the ducts coming from the liver to form the *common bile duct*. The structure of the gallbladder and biliary system looks like this:

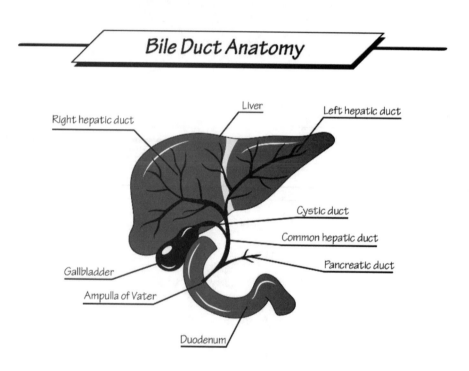

Bile exits the liver and gets in through the cystic duct. This leads directly to the gallbladder, where the bile gets concentrated. After a meal, the gallbladder contracts. Bile goes back down the cystic duct and enters the *common hepatic duct*. Pancreatic enzymes get added later through a common duct that will enter the duodenum through a tiny hole called the *ampulla of Vater*.

Fun Factoid: *The ampulla of Vater is the spot where doctors insert the dye into the biliary tract in the endoscopic retrograde cholangiopancreatography discussed earlier. The entire biliary tract will light up on an X-ray when the dye is pushed into it. This is often where bile duct stones can get stuck. Some people need to have a small tube called a stent placed in the ampulla of Vater to keep the duct open all the time.*

The Exocrine Pancreas

The pancreas is a gland with a major role in both the endocrine and digestive systems. It is a long and thin organ located behind the stomach inside the abdominal cavity. It is about 6 inches in length. The digestive part of the pancreas is called the *exocrine pancreas*, while the endocrine part of the pancreas is called the *endocrine pancreas*. They have two completely separate jobs, but you can't tell which part is which just by looking at it. The two functions are mixed within the entire pancreatic structure and can only be identified under the microscope.

The exocrine pancreas secretes a pancreatic juice that contains digestive enzymes, which are responsible for much of the GI tract's ability to absorb nutrients. These enzymes are used to break down the lipids, proteins, and carbohydrates in the partially digested food that arrives at the duodenum from the stomach.

Anatomically, the pancreas is divided into the head, the neck, the body and the tail. The head is surrounded by the duodenum and surrounds two blood vessels, the superior mesenteric artery and the *superior mesenteric vein*. The neck of the pancreas is about 2.5 centimeters long and sits between the body and head, in front of the superior mesenteric vein and artery. Its front upper surface supports the pylorus, or base of the stomach.

The neck of the pancreas arises from the left upper aspect of the front of the head. It is directed upward and forward at first, and then goes upward and to the left to join with the body of the pancreas. The body is the biggest part of the pancreas. It sits behind the pylorus of the stomach at the same level as the *transpyloric plane*. The tail is on the far left-hand side of the abdomen next to the spleen. This is what the pancreas looks like:

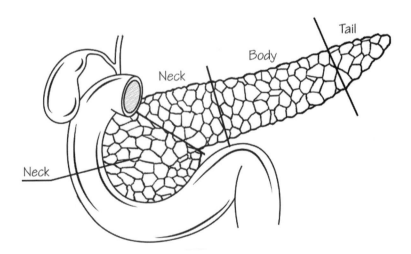

The exocrine pancreas has two main ducts that drain digestive enzymes. These are the *accessory pancreatic duct* and the *main pancreatic duct*. They drain the digestive enzymes through the ampulla of Vater and into the duodenum.

If you look at the pancreas under a microscope, you'll see that there is a clearer division between the endocrine pancreas and the exocrine pancreas. When the pancreatic tissue is stained with dye, clusters of endocrine cells called the *islets of Langerhans* or the *pancreatic islets* are visible. There are also darker staining cells known as *acini*, which are arranged in lobes that are separated from one another by a thin fibrous barrier. These acini are part of the exocrine pancreas because they secrete pancreatic enzymes.

The main difference between an exocrine gland and an endocrine gland is the presence of a duct system in exocrine tissue. Both types of tissue make some type of substance that acts on other body cells and tissues but, with exocrine glands, the substance goes through a duct to act on those tissues supplied by the duct. With endocrine glands, the substances made by these glands are called *hormones*. These hormones are released into the bloodstream, where they can act on cells or tissues far removed from the gland itself.

The secretory cells of each acinus surround a small intercalated duct. The intercalated ducts then drain into bigger ducts within the lobule, and finally become interlobular ducts. These all come together to form one of the two drainage ducts called *pancreatic ducts* leading out of the pancreas. They often join with the common bile duct so that everything dumps into the duodenum together.

How Food is Transformed into Nutrients

The digestive system, like the other systems of the body we've discussed, is a well-oiled machine as long as everything works as intended. It sounds simple to think that all you have to do is eat and the rest takes care of itself. For a healthy person, it does take care of itself, but it also involves an integrated interplay between organs of the digestive system. Food is taken in, mechanically and chemically digested, and absorbed by the intestinal tract. Lastly, there are mechanisms that get rid of wastes we cannot utilize.

In this section, we will talk about how the food you eat is manipulated by the organ systems of the digestive tract to make the most use out of every bite. As you'll see, it starts with chewing and swallowing effectively, and goes on to become a complex process that ends with nutrients arriving in usable form to the tissues in order to be metabolized on a cellular level.

Digestion in the Stomach

The opening of the stomach where food must pass to enter it is called the *LES*, or *lower esophageal sphincter*. This sphincter should close again after the food bolus has passed; but if you've ever had heartburn, you know it doesn't always do this. If you eat foods that are fatty or acidic, or if you simply eat too much food at once, you're increasing the chance that the food or just stomach acid will back up into the esophagus, which is commonly called *heartburn*. Drinking alcohol, being obese, consuming caffeine, and smoking will also contribute to heartburn.

Once in the stomach, both mechanical and chemical digestion take place. The stomach muscles churn in order to mix the food as it sits in the stomach. Chemical digestion happens when cells of the stomach make hydrochloric acid, which denatures proteins and breaks down food purely by the fact that the stomach environment is so acidic. The stomach makes a thick mucus layer that coats the lining in order to protect the lining itself from getting digested or damaged by the action of this acid.

The main stomach enzyme is called *pepsin*. Its primary goal is to break down proteins into small *oligopeptides*, or short peptide chains. It is secreted as a precursor enzyme called *pepsinogen* by the chief cells of the stomach. Once exposed to acid, pepsinogen becomes activated as pepsin in order to act on the proteins you have eaten.

While peristalsis mixes food in a churning kind of way, the general idea is to use peristalsis to push food through the stomach toward its most distal end. As the food gets to the pylorus and finally exits through the pyloric sphincter, it is finely churned and partially digested. The pyloric sphincter will open periodically in order to allow just a little chyme through at a time; the idea is to avoid sending too much of the chyme substance into the duodenum at once.

The Role of the Small Intestine

The small intestine starts with the duodenum, which receives small amounts of chyme at a time from the stomach. This chyme is very acidic, which is a problem because the duodenum doesn't have the same protective structures in place to protect itself from acidic substances. Furthermore, the enzymes necessary to further digest food in the small intestine do not work well under acidic circumstances.

The first thing that happens is that bicarbonate is released into the proximal duodenum. Bicarbonate is alkaline rather than acidic, so it counteracts the acidic hydrochloric acid coming from the stomach, bringing the pH of the duodenum's lumen up to a level that allows pancreatic enzymes to function.

Pancreatic enzymes have three major functions: to break down carbohydrates, fats and proteins. Proteins, peptides and amino acids are digested in the small intestine by enzymes secreted by the pancreas designed exclusively to break up peptide chains. These main enzymes include *trypsin* and *chymotrypsin*. The goal of these enzymes is to continue the process started in the stomach, which is to break longer peptide chains into oligopeptides and amino acids that can be absorbed into the intestinal wall.

Other enzymes, such as those called *lipases*, are secreted by the pancreas in order to break down fats in the diet. Lipases break down the triglycerides in the food into free fatty acids and monoglycerides. The whole process is aided by bile salts that have been secreted by the liver and stored in the gallbladder.

This is how bile salts and lipases work together to break down lipids. The lipases are soluble in water but the triglycerides aren't. For this reason, bile salts bind with these fats and emulsify them in the same way that detergent emulsifies the grease in your sink when you do dishes. They form a structure of fat coated with bile salts that holds onto the triglycerides in an environment where the lipase can break them down into fatty acids. Those fatty acids can then be absorbed by the intestinal villi in the rest of the small intestine. This is what it looks like:

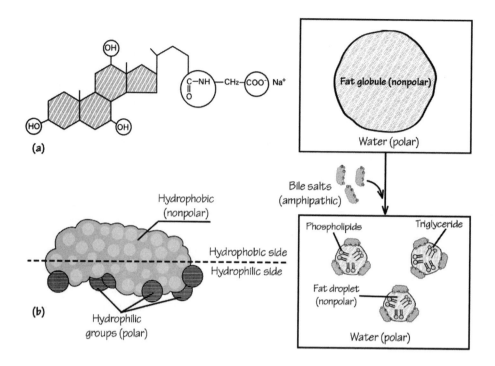

Carbohydrates are broken down into simple sugars and *monosaccharides* such as glucose. Pancreatic amylase breaks down some carbohydrates into *oligosaccharides* as well. Some carbohydrates and fibers pass undigested into the large intestine where they may, depending on their type, be further digested by intestinal bacteria.

You might think that after all of these enzymes have done their job, the whole process might be enough to break food down enough to be fully absorbed by the enterocytes of the small intestine. These are very small particles, certainly, but they aren't always small enough. This is where the small intestine itself takes over. Along the mucosa of the small intestine are the lining cells with villi and even tinier microvilli on top, which jut into the lumen of the alimentary tract. This rough layer is called the *brush border*.

The enterocytes and brush border look like this:

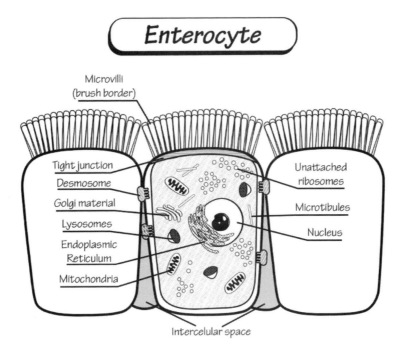

Enterocyte

Microvilli
(brush border)

Tight junction
Desmosome
Golgi material
Lysosomes
Endoplasmic
Reticulum
Mitochondria

Unattached
ribosomes

Microtibules

Nucleus

Intercelular space

Right at this level and closer to these enterocytes, the cells themselves make a whole host of brush border enzymes that further break everything down into the smallest possible absorbable unit—namely, fatty acids, amino acids and simple sugars. These easily get absorbed into the enterocytes, passing through them and into the circulation surrounding the intestinal wall. From there, most of these go directly to the liver, except for fatty acids, which instead enter the lymph system first in order to go into the circulation through a different path.

The Role of the Large Intestine

The large intestine functions in the absorption of water and any remaining nutrients that haven't been digested by the small intestine. It absorbs vitamins that are made by the colonic bacteria, such as vitamin K. (This is crucial because humans do not get enough vitamin K in their diet.) Vitamin B12, riboflavin and thiamine are also absorbed by the large intestine. The colon is full of healthy bacteria that help to make or process many of these vitamins.

The bacteria inside the colon also break down some of the undigested fiber for their own metabolism and create *butyrate*, *propionate* and *acetate* as waste products. These in turn are used by the lumen cells lining the colon for nourishment. About 10 percent of the undigested carbs is available for the bacteria, although this can vary with what you eat. Foods high in fiber like raw vegetables and whole grains aren't digested easily by humans, but our gut bacteria love them. The colon makes no digestive enzymes. All chemical digestion happens in the small intestine prior to the chyme reaching the large intestine.

While we talk about water reabsorption being the job of the large intestine, this is partly a myth. In actuality, about 80 percent of water is absorbed by the small intestine, with 10 percent absorbed by the colon. The last 10 percent is excreted along with feces. Even so, you need the large intestine to regulate the last bit of water absorption in the stool in order to make it the right consistency for easy passage. If you don't drink enough, even the good work of the intestinal tract in regulating water reabsorption won't be enough to spare you from being constipated.

Fun Factoid: What's in a fart? It turns out that you make up to a liter and a half of intestinal gas every day (but not all at once). 99 percent of this gas has no odor whatsoever. These gases include methane, hydrogen and carbon dioxide, much of which is either swallowed or made by your intestinal bacteria. If some of this gas is instead hydrogen sulfide, which is a sulfur compound, it'll result in a fart you can smell.

Another function of the colon is the compaction of feces, and the storage of fecal material in the sigmoid colon and rectum before it can be passed through the anus. There are sphincters you're in control over that will open upon command when the rectum signals it's full. Through large peristaltic contractions of the colon and the action of your stomach muscles, you are able to pass stool of varying consistencies through the anal canal and into the toilet.

The GI Tract

One unsung role of the GI tract is that it makes gastrointestinal or gut hormones. There are so many of them that we aren't completely sure of the total number or exactly what they do. Some of these are involved directly in the process of food and eating. Two of them are called *ghrelin* and *leptin*.

Ghrelin is the hunger hormone made by the liver and stomach. Think "gremlin" and you'll be able to remember that it makes you as hungry as a gremlin. It tells your brain you're hungry and stimulates your appetite. Opposing this hormone is leptin, which is made by fat cells. It doesn't necessarily keep you from eating, but over time will regulate how much food you take in. There are some drugs used to fight loss of appetite in cancer patients by mimicking ghrelin, while drug companies are busy looking for similar drugs to antagonize the hormone in people who are obese.

Research studies on some of these hormones have shown that many of the gut peptides, such as *substance P*, *secretin* and *cholecystokinin*, seem to play an important role as *neuromodulators* and *neurotransmitters* in the peripheral and central nervous systems. Their role is complex since the gut and the brain are considered so closely connected. The neuroendocrine cells that make these hormones don't form specific glands, but are spread throughout the GI tract.

Gastrin is a peptide hormone made in the stomach that stimulates the secretion of gastric acid by the parietal cells of the stomach. It is made by G cells in the pyloric antrum of the duodenum, stomach and pancreas. Ultimately, it binds to receptors on the parietal cells of the stomach, activating a *potassium-hydrogen ion ATPase pump* in the apical (internal) membrane of these cells. The end result is an increase in hydrogen ions in the stomach cavity. The release of gastrin is stimulated by peptides inside the stomach. This is exactly what you want, because the job of hydrochloric acid is to break these peptides down.

To Sum Things Up

The gastrointestinal system is responsible for taking in the food you eat and mechanically and chemically digesting it so that it's broken down into its most basic components. Once this has happened, these simple molecules are absorbed by enterocytes in the small intestine and sent into the body for cellular metabolism.

While it sounds like a simple process, it actually involves the coordination of many voluntary and involuntary actions. Taking in food and chewing it is entirely voluntary; but after you start the swallowing reflex, the process of peristalsis and the digestion of food is generally involuntary. These functions rely on the activity of the parasympathetic nervous system and the vagus nerve. After digestion comes the absorption of nutrients, which usually happens through the activity of the small intestinal cells lining this structure.

Food that isn't easily absorbed, such as fiber from fruits and vegetables, gets partially metabolized by gut bacteria, while the rest gets consolidated as stool that you're able to pass voluntarily, using the involuntary muscles of the colon and your own voluntary abdominal muscles.

Chapter Eleven: Your Nervous System and Special Senses

Your nervous system is so essential to life that you would not be conscious, able to perceive your environment, or able to move without it. On a cellular level, nearly every cell of your body has some kind of nerve supply. Your nerves help to direct how much blood each cell gets and how active the metabolism of the cell should be.

This is an incredibly intricate system that involves not only your brain but an interconnected highway of nerves that orchestrate every process in your body to some degree. In this chapter, we will talk about the central nervous system, which is basically the brain and spinal cord, the peripheral nervous system, and the subconscious autonomic nervous system.

Nervous System Basics Explained

Your nervous system is much more than the part of your body that brings together the functions of movement and sensation with the neurons of your thinking brain. You have parts of your nervous system that don't ever come in contact with your brain, and there are certainly parts of your thinking brain, or your *cerebral cortex*, that never have connections to the rest of your body.

This leads to a very complicated system of control over your body and your experiences, that goes way beyond wanting to pick up a can of soda and having your arm obey that command. Sometimes, for example, you are so thirsty that the non-thinking parts of your brain send signals that lead to you pick up the soda and drink it without even having a solid conscious thought. If you had to think first and act only on a given thought every time you moved, you would be so wrapped up in that type of dynamic, you wouldn't be able to get anything else done.

Fortunately, your nervous system doesn't work like that. In a large corporation, which is a lot like your functioning nervous system, there are people at the top like the CEO, and workers at the bottom who do the everyday jobs from deep within the corporation. It would be silly for the CEO to send a memo to the clerk in accounting every time the clerk needed to do something in his job.

Instead, the workers do the jobs they've been trained to do, and the CEO knows nothing about these activities unless there's some kind of screwup that's significant enough to get back to the top. Yes, the whole corporation works toward the same end, but it is rarely completely a top-down experience. Sometimes, it all works in reverse and becomes a bottom-up experience, where the clerk in accounting sends an emergency memo to the CEO saying that the budget is messed up, or that the company is out of funds due to a computer glitch.

There are CEOs, or thinking neurons, in the brain, but there are also working nerves in your spinal cord, sensory receptors in your skin, and *special senses* that are constantly surveying the environment for you, so that they can be aware of the things that matter.

Structurally speaking, we often think of the nervous system as being divided into the central nervous system and the peripheral nervous system. Functionally, however, things are more complex than that. You can divide the nervous system functionally in several different ways that include your *somatosensory nervous system*, your motor nervous system, the autonomic or visceral nervous system, and the *enteric nervous system*. These functional parts of the nervous system are still housed somewhere in both the central and peripheral nervous system compartments.

The central nervous system is basically just the spinal cord and the brain. The peripheral nervous system is defined as everything outside of the brain or central nervous system. In the peripheral nervous system, there are numerous highways of nerves that are nothing but encased bundles of axons, which are the long fibers in your nerves coming off the main nerve cell body.

In your central and peripheral nervous system, it helps to think about everything being a two-way street. Nerves that leave the central nervous system in some way are called *motor nerves* or *efferent nerves*. There are also incoming nerve fibers that enter the central nervous system in some way. These are called *sensory nerves* or *afferent nerves*. Between these are *interneurons* that never leave the central nervous system but act as connecting nerves, bridging any gaps that might exist between the central and peripheral parts of your nervous system.

In reality, when you look at any individual nerve in your body outside of the spinal cord, few of these are purely motor nerve fibers or purely sensory fibers. Almost all of them are two-lane highways, with both sensory and motor nerve fibers making use of the same physical space.

Nerves that leave the cranium directly and never reach the spinal cord are known as *cranial nerves*, while those that exit from the spinal cord are known as *spinal nerves*. Cranial nerves usually control activity of the muscles and sensations of the face and neck, but at least one of these nerves wanders as far down as your digestive tract. Spinal nerves usually have a direct connection to the spinal cord and control the functions of your arms, legs and trunk.

We will soon talk more about the basic cells of the nervous system. These are divided into the neurons, which do all of the communicating within the nervous system, and the *neuroglia*, or glial cells, which are not electrically active but support the neurons in some way. You can think of neuroglia as helpers who act just like the janitors and housekeepers in the corporation. They make no major decisions, but the whole system would quickly fall apart if they didn't do their job to keep the whole company functioning smoothly.

Neurons and How They Work

Your neurons are the major electrical cells of the nervous system. These are the cells that communicate with each other in order to coordinate everything the nervous system needs to do. Like the heart muscle cells we've talked about, these are cells that can depolarize, with the ability to send the electrochemical signal from one end of the cell to another. The only real difference in how this works is that, while heart muscle cells depolarize on their own in a rhythmic fashion, neurons are more sophisticated and only fire off a signal when they receive a message from another neuron.

The basic neuron looks like this:

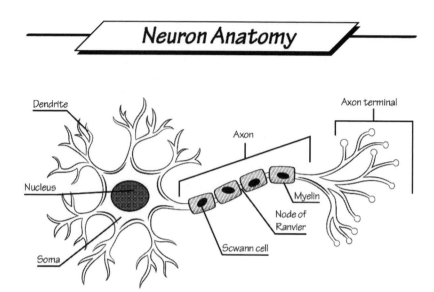

If you look at this image carefully, you'll see that there are several parts to these unique cells. Each neuron has a cell body, called a *soma*. This is where the cell nucleus is and where most of its other cellular organelles are found.

You can also see several extensions or projections from the cell body. These are called *dendrites* and *axons*. There are often multiple shorter dendrites and just one long axon, but this isn't always the case. In general, dendrites are the receivers that gather information from other nerve cells, while axons are the communicators that send signals outward from the nerve cell body, to deliver information to the next nerve cell down the line (or to a muscle cell of some kind).

These are general guidelines, though. It's possible for a dendrite to be a delivering part of the neuron and to send a signal to another dendrite, axon, or even the cell body of another nerve. It is also possible to have longer dendrites and shorter axons. One general rule of thumb, though, is that many (but not all) axons are *myelinated*, meaning that the axon is coated with a fatty substance called *myelin* that allows the electrical signal to pass along the axon at a much faster rate.

Myelin is made by certain glial cells. It does not exist as a long tube of myelin down the whole axon, but is attached in clumps along the axon with small spaces between each batch called the *nodes of Ranvier*. As we talk about how nerves work, you will see why myelin is such a good thing for nerve cell transmission, and what these nodes of Ranvier do.

Nerve cells are not generally wired together like the wiring of your house. Instead they come very close to one another, and pass on their message chemically though spaces called *synapses*. An axon might have an electrical signal pass down it until it reaches the end. This causes chemical and enzymatic reactions to occur at the axon terminal. From there, the message is sent on by delivering a chemical called a *neurotransmitter* into the *synaptic cleft*, which is what the physical space between the neuron is called.

The next neuron in line (called the *postsynaptic neuron*) receives the chemical message through receptors on its dendrite surface and depolarizes, turning the process back into an electrical one. This electrical signal then speeds on through the postsynaptic cell. This is basically what a synapse looks like between two nerve cells:

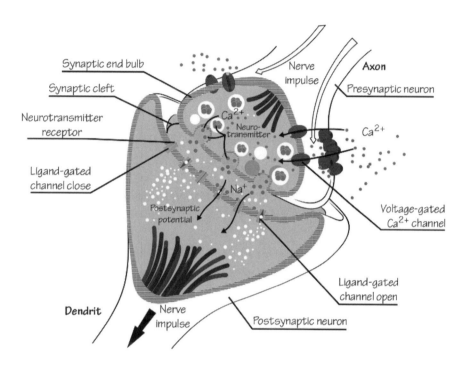

Not all synapses happen between neurons. The neuromuscular junction in the muscles is a synapse that connects a motor neuron to a muscle cell. There is a neurotransmitter in these synapses called *acetylcholine* that takes the motor neuron's signal and transfers it to the muscle cell, telling the muscle cell to contract. There are also synapses between nerve cells and glands that help to decide how the gland will behave.

You can see how a single neuron never acts by itself. Each neuron has the potential to connect with hundreds and thousands of other neurons or other cells in the body. It can become incredibly intricate—even more complex than sending telephone messages. In fact, in the cerebellum, one Purkinje cell (which is a specialized nerve fiber) receives some kind of input from as many as 200,000 other neurons.

Neuroglia, or Supporting Cells

Glial cells have many supportive functions in the nervous system. There are several different types of glial cells that act in unique ways to make sure the nervous system works properly.

Glial cells are commonly known as the glue of the nervous system; however, this is not completely accurate. Glia was first discovered in 1856 by the pathologist Rudolph Virchow in his search for a connective tissue in the brain. Through his work and others, we can currently identify four main functions of glial cells:

- To surround neurons and hold them in place
- To supply nutrients and oxygen to neurons

- To differentiate one neuron from another
- To destroy pathogens and remove dead neurons

Let's look at these neuroglial cells and see what they do:

- **Oligodendrocytes:** These are found only in the central nervous system. They surround some of the axons in order to secrete myelin, a fatty insulating substance that increases the speed of the electrical signal traveling in these areas.
- **Astrocytes:** These are called *astrocytes* because of their star shape. The points of these star-shaped cells extend out to different neurons in order to supply nutrients to the nerve cell and support them in many general ways.
- **Ependymal cells:** These are cells that make the cerebrospinal fluid that bathes the brain and spinal cord surfaces. They line the internal ventricles of the brain. The cerebrospinal fluid these cells make provides nutrients and remove wastes from the central nervous system.
- **Microglia:** These are the cells of the central nervous system that survey the area for debris and potential pathogens, acting as immune cells to clean out areas of the nervous system that need this kind of servicing.
- **Satellite cells:** These are similar to astrocytes but act only in the peripheral nervous system. They wrap around the nerve cell body in order to support the nerve cells structurally and nutritionally.
- **Schwann cells:** These are supporting cells similar to the oligodendrocytes that make myelin in order to speed transmission along myelinated axons. The big difference, though, is that these cells act only in the peripheral nervous system, essentially doing the same job as the oligodendrocytes but in the periphery instead of the brain or spinal cord.

While glial cells are not necessarily electrically active, the nervous system can't function properly without them. There are some diseases, for example, that involve loss or damage to the myelin sheath around axons. This leads to slow transmission of nerve impulses throughout the nervous system. The most common disease is multiple sclerosis.

Fun Factoid: People with multiple sclerosis have very little myelin around their axons, leading to slowed nerve fiber transmission. This affects all parts of the nervous system where myelin is necessary. It is believed to be an autoimmune disease, where the person makes antibodies or some other immune-related response that destroys their own myelin sheath. Drugs used to block the immune system can be effective in treating this disorder, even though there is no known cure for it.

How Neurotransmission Happens

Neurotransmitters come in different types. Some are small peptides or short proteins, such as substance P, which is often a neurotransmitter involved in pain signaling. Others are amino acids like *glutamate*. *Biogenic amines* are made from amino acids but are not themselves amino acids. *Dopamine, serotonin* and *norepinephrine* all fall into this category.

Fun Factoid: Have you ever heard of capsaicin cream? It's a common pain cream essentially made from the same substance that makes hot peppers so hot. It acts on pain by decreasing the amount of Substance P in the nerve fibers near the skin. This decreases the signaling of the sense of pain to the

Not all neurotransmitters automatically tell the postsynaptic cell to depolarize enough to send the signal onward to other cells. There are also *excitatory neurotransmitters* and *inhibitory neurotransmitters*. Most can be either excitatory or inhibitory, depending on how the message is received by the next cell in line.

Excitatory neurotransmitters include one called *glutamate*, which is almost always excitatory. When glutamate is released into the synapse, there are receptors on the postsynaptic cell (usually at the dendrite) that cause this postsynaptic cell to depolarize or change the membrane potential across the cell membrane. This is the electrical signal necessary for the postsynaptic cell to pass the signal on even further to still other cells.

Inhibitory neurotransmitters include one called *GABA*, or *gamma aminobutyric acid*, which is almost always inhibitory. If a presynaptic neuron releases GABA into the synaptic cleft, the GABA molecules bind to its own postsynaptic neurons. The big difference, though, is that when these inhibitory neurotransmitters attach to the receptor, it causes the reverse of depolarization, called *hyperpolarization*. When this happens instead, the postsynaptic neuron will not pass on a signal to any other cell. It's basically a "stop" message that discourages further activity by the postsynaptic cell.

So, what happens if a nerve cell gets mixed messages? This actually happens all the time. It would be like having some people in a crowd urging a person to jump out of a six-story building, while others tell that same person not to jump. The person might be more inclined to take the leap if they hear more jump messages than the contrary. It would be sort of a democratic decision based on how much of each message the person hears.

In some cases, this is exactly how the postsynaptic cell functions. In a democratic fashion, the nerve either depolarizes or doesn't based on how loud the collective voices are, or which message is the loudest at any point in time. In other cases, the postsynaptic cell might only listen to the last voice it hears.

The act of jumping from a six-story building is truly an all-or-nothing event, that can't be taken back or done halfway once the decision has been made and acted upon. The nerve cell is exactly the same way. It will either depolarize and fire up a new signal for the next cell in line, or it won't depolarize at all and there will be no signal involved. The part of the process that isn't all-or-none is the input the neuron receives. If the input says "fire" overall, the postsynaptic cell will do this; if the input says "don't fire" overall, the postsynaptic cell will behave and won't do it.

Of course, a neurotransmitter cannot send out a signal and just stay like that, continually stimulating or inhibiting a postsynaptic cell indefinitely. The signal has to stop somehow or the postsynaptic nerve cell would just keep firing. The nervous system just doesn't work this way and must constantly adapt and change with the different circumstances. In order to do this, there must be a way for the neurotransmitter message to be temporary.

This is accomplished in several unique ways. In some cases, the neurotransmitter simply diffuses away from the synaptic cleft over time, so that the signal is no longer strong enough to depolarize the postsynaptic cell. In other cases, there are specific enzymes in the synaptic cleft that break down the neurotransmitter as soon as the message is passed on. This would be like having a continual paper-shredding machine that destroys a message as soon as it's read. Lastly, there are mechanisms in place that will collect the neurotransmitter as soon as it does its job, taking it back up into the presynaptic cell so it can be reused again.

Fun Factoid: Some of the more commonly used drugs for depression are called selective serotonin reuptake inhibitors, *or* SSRIs. *These drugs act on the neurotransmitter called* serotonin, *which doesn't seem to be as active as it should be in depressed people. SSRIs work by blocking the reuptake of serotonin in the synapse so that when it's released, it cannot easily get back into the presynaptic cell. This makes its effect on the postsynaptic cell last longer, which improves depressive symptoms.*

Nerve cells are able to form elaborate networks through which nerve impulses travel. Each neuron has about 15,000 connections with other neurons. You can see how complex and messy it can get if everything isn't well-coordinated. Fortunately, there are parts of the nervous system that have specific coordination functions that allow nerve cells and their connections to stay in order.

The Central Nervous System

The central nervous system (CNS) consists of the brain and the spinal cord. This is not an arbitrary designation but is based on the fact that these two areas of the nervous system are protected from the rest of your body by the *blood-brain barrier*. This is a protective endothelial cell layer unique from the rest of the body. It's only found in the parts of the central nervous system that need to be isolated in some way.

The cells that form the blood-brain barrier are epithelial cells in the capillary walls of the brain and spinal cord that have *tight junctions* wedged between each cell. Tight junctions are exactly what they sound like—they form a tight barrier between the capillary endothelial cells, in order to prevent most things from gaining access to this fragile part of your nervous system. Only lipid-soluble molecules, small molecules, and certain gases are allowed to pass through this layer.

Fun Factoid: Why does alcohol make you drunk? It's because alcohol is one of the few substances that can cross the blood-brain barrier. A drug or substance like alcohol can only cause central nervous system effects if it can cross the blood-brain barrier. CBD oil is part of the marijuana plant that doesn't make you high, even though it crosses the blood-brain barrier. This is because it only binds loosely to the receptor that makes you high, and when it does, it blocks it instead of activating it.

While the skull and other structures protect the central nervous system from physical damage of some kind, the blood-brain barrier has its own protective function in preventing CNS damage from pathogens like bacteria, viruses and toxic substances. Without this barrier, the brain would be too vulnerable and would have a much greater chance of damage.

The Brain Is a Regulating Organ

The brain is the central processing center of your nervous system. If your nervous system was some kind of corporation, the brain would be the CEO and top board members. A lot of cognitive or thinking processes happen in the brain. It's where you make sense of your environment, store memories, process emotions, and make decisions about what intentional body movements you're going to make.

Structurally, your brain has many different parts, made up of 15 to 33 billion neurons that generally do not regenerate themselves if they are damaged or die. Your cerebral cortex is where major thought processes happen. It is divided into two *hemispheres*—left and right—that are connected to one another through a large central bundle of nerves called the *corpus callosum*.

The brain surface is convoluted, with ridges and valleys. The ridges are called *gyri* (singular is gyrus) and the valleys are called *sulci* (singular is sulcus). These will look slightly different from person to person and have different functions in various parts of the brain. For example, the *somatosensory cortex* is located in the *parietal lobe* of the brain, in a gyrus just behind a major sulcus called the *central sulcus*.

There are four lobes in the cerebral cortex. Here's what they look like:

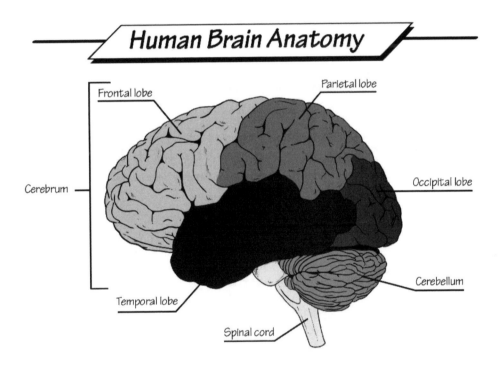

Each lobe of the cerebral cortex has different functions. Here is a brief rundown of what each one is responsible for:

- **Frontal lobe:** This part of the brain has the motor cortex, which controls planned movement. It also contains important parts that help in defining your personality and helping you with judgment, planning, and control over your emotions. One of the two major language centers called *Broca's area* is located here.
- **Temporal lobe:** This has an important role in auditory processing and contains *Wernicke's area*, which helps you give language its meaning. Deeper structures of this lobe are involved in your *limbic system*, which is a major emotional center in the brain.

- **Parietal lobe:** This part of the brain has the somatosensory cortex, which is where you process the different senses coming in from the peripheral sensory receptors and sensory nerves of the body.
- **Occipital lobe:** This is where your *visual cortex* is located. Images from your eyes must travel all the way to this back part of the brain in order to be processed and interpreted.

You may have heard of the terms *white matter* and *gray matter* when referring to the brain. They are named for their color, of course. Gray matter is where neuron somas are located. They can exist in clusters called *nuclei*, which include neuron cell bodies that share the same function. They also exist along the outer surface of the cerebral cortex.

The white matter is white because of the myelin surrounding the axons. Wherever you see white matter, you know you are looking at myelinated nerve axons rather than nerve cell bodies (which are never myelinated). White matter contains tracts where axons are traveling from one part of the brain to another. In the cortex, the white matter is deeper than the gray matter and represents places where the neurons are busy sending a signal through a nerve tract. There are no real nerves in the brain like you see in the body, only these tracts that involve similar nerve axons traveling in the same physical space.

There are many deeper structures in the brain that aren't necessarily a part of the cerebral cortex. Some of these structures are depicted in this longitudinal section of the middle of the brain:

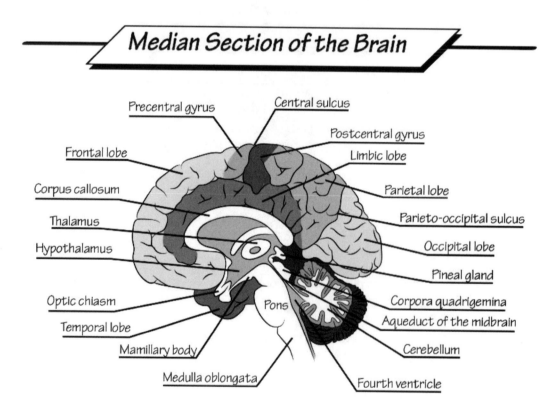

Let's look at these deep brain structures and learn a little bit about what they do:

- **Thalamus:** This is a deep brain structure that acts like a relay center for sensory information in the brain. It takes all incoming sensations except those from your sense of smell, and relays these signals to all parts of the brain.
- **Hypothalamus:** This is a small structure in the midbrain area that has the big job of participating in homeostasis of the body. It regulates the pituitary hormones in the endocrine system and helps to maintain normal body temperature, appetite, and many subconscious processes involved in the autonomic nervous system.
- **Hippocampus:** This is a midbrain structure responsible for managing memories, and for transferring your short-term memories into long-term ones. It is also a part of your emotional limbic system, which is why your emotions are so closely linked to your ability to remember things and to past memories you have.
- **Amygdala:** This is a deep temporal lobe structure where you process fear. Images and sensations get interpreted in the amygdala as being fearful or not, depending on your emotional state and perceptions. It connects with other structures in your limbic system.
- **Cerebellum:** This basically means "little brain" and is located in the back and inferior part of your cranium. It is involved with your ability to have smooth movements and normal balance and posture, as you participate in the different subconscious motor activities you do all the time.
- **Pons and medulla:** These are collectively called your *brainstem*. All the nerve tracts to and from the spinal cord pass through these parts of the brain. There are also multiple nuclei that control subconscious processes like respiration, heart rate, blood pressure and level of alertness. These are the oldest parts of your brain from an evolutionary perspective, and involve largely subconscious processes in your nervous system.

The Spinal Cord

Your spinal cord is a continuation downward of the medulla oblongata in the brainstem. It contains multiple tracts and nuclei that are crammed together into a superhighway, that runs from the cervical area in your neck to about the L2 level of your spine. Look at the following image of your spine. Notice the tiny holes you can see best in the thoracic and lumbar areas on either side of the spinal column. These are the intervertebral foramina, which are very important bony canals on either side of the spine where spinal nerves travel.

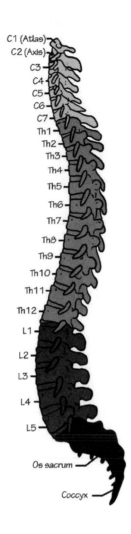

The job of the spinal cord is to send signals from the brain to the peripheral nerves, and to take peripheral sensory information from the body back up to the brain. The intervertebral foramina are the places where the central nervous system ends and the peripheral nervous system begins. There are 31 spinal nerves on either side of the spinal cord that come off at the different levels to supply segments of the body at every level.

Each spinal nerve has motor and sensory parts, and each serves its own section of the body called a *dermatome*. This figure shows you what the different dermatomes look like in the body:

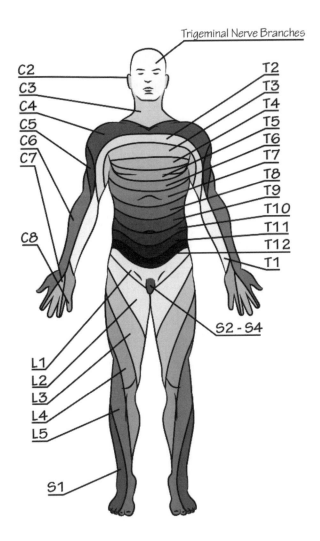

When we talk about the sensory and motor systems, you will see that the spinal cord plays a role in both of these. As a general rule, the back or dorsal part of the spinal cord is most concerned with sensation and sensory pathways, while the front or ventral part of the spinal cord is most linked to motor nuclei and motor nerve tracts. We will talk in a minute about how the spinal cord participates in reflexes, such as the knee-jerk reflex, which you can see consciously but cannot consciously control.

Sensory and Motor Pathways

Your somatosensory and motor systems are separate pathways that are, of course, linked together so that you can sense things in your environment, process them in the brain, and make motor decisions about what you perceive. Let's start with the somatosensory part of this first.

Senses: Body to Brain

There are sensory fibers coming in from the body all the time, providing information from receptors in your skin, subcutaneous tissues and viscera. There are receptors for pain, light touch, vibration, temperature, and pressure or stretch of the skin that we will talk about in a little bit. There are also proprioceptors in the joints and moving tissues of the body (like your muscles and tendons) that help you determine where your body parts are in space, even without looking at them directly.

There are incoming signals that travel through nerves to the spinal cord, where they form tracts that ascend up to the brain along the dorsal aspect of the spinal cord. The dorsal column is where these fibers mainly travel, although there are some fibers that travel along a different path, called the *spinothalamic tract*. These two tracts are visible in this image:

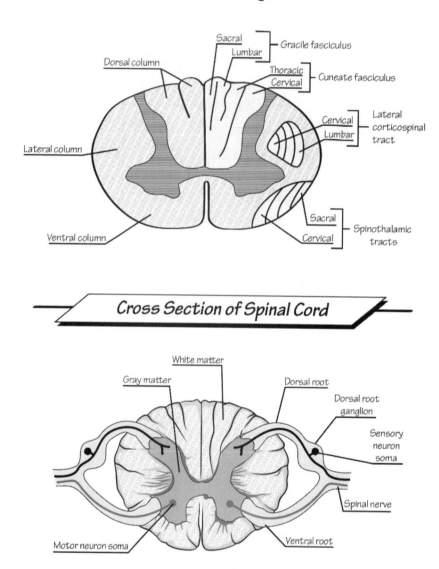

The nerve fiber from the arm, for example, will sense light touch and will send a signal through a nerve cell body in a *dorsal root ganglion* along the side of the spine. A ganglion is the same thing as a nucleus but is located outside the central nervous system. It's still a collection of nerve cell bodies, but is just not located in the brain itself.

There's a connection made in the dorsal root ganglion, which sends its own axons up the dorsal column, where they end in the medulla oblongata in several brainstem nuclei. Here there is another neuron that sends fibers up to the *thalamus*. Before doing this, though, the axon fibers cross over, or *decussate*, to the opposite side of the brainstem. This is why your right brain processes information it gets from the left side of your body.

The nerve axons that instead go up the spinal cord through the spinothalamic tract are those that sense pain and temperature. These decussate in the spinal cord at different levels before traveling directly up to the thalamus. The thalamus has many of its own nuclei and takes all this information, sending it further on to the somatosensory cortex of the parietal lobe where you actually perceive the feeling you felt in your arm.

The somatosensory cortex is amazing. It has different parts that process all the information from the periphery of the body in unique ways. There is, for example, more of this cortex that is devoted to your lips and fingers than there is of your back. That's because the lips and fingers have a lot of sensory receptors and are sensitive parts of your body. The sensory homunculus is a funny depiction of the way this somatosensory cortex in the brain is arranged:

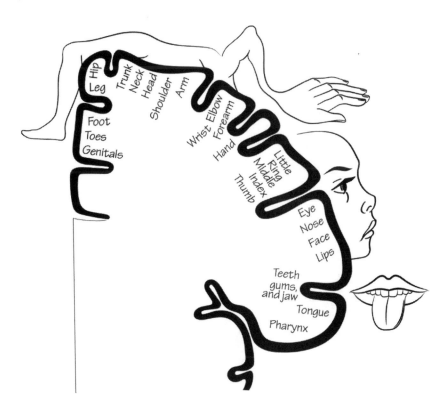

Motor Pathways: Brain to Peripheral Nerves

Just in front of the somatosensory cortex is a matching motor cortex called the *primary motor cortex* in the back part of the frontal lobe, separated from the somatosensory cortex by the central sulcus. It's probably no accident that these two areas are so close to each other, and why the motor homunculus looks basically the same as the sensory one. It's in this primary motor cortex that you first think about moving your arm, for example.

Say you decide to move your arm, because your senses tell you a fly has landed on it and it bothers you. Assuming you pay attention to the whole thing and don't subconsciously brush off the fly, you have *Betz cells* in your cortex that initiate movement. This starts the downward or descending pathways from these Betz cells to your peripheral motor nerves.

The primary motor cortex does not work by itself. There is input from secondary motor centers of the brain and cerebellum, so that movements can be coordinated. The Betz cells are large neurons that send axons through two tracts, the *corticospinal tract* and the *corticobulbar tract*. The corticospinal tract contains axons destined for the spinal cord and for movement to be controlled by the body's muscles, while the corticobulbar tract contains axons that do the same thing for the cranial nerves that supply the face and neck muscle through the cranial nerves.

The cells from the cortex to the spinal cord are called *upper motor neurons*. They travel down a white matter pathway through a bundle called the *internal capsule*. In the medulla, these neurons can be found in a larger bundle area called the *pyramids*. This is where the motor fibers cross over. Interestingly, not every fiber crosses over. Some will descend on the same side of the body so that you can better control your posture, which relies on coordination between the right and left side of the body at the same time. This image shows these major pathways in the brain and spinal cord:

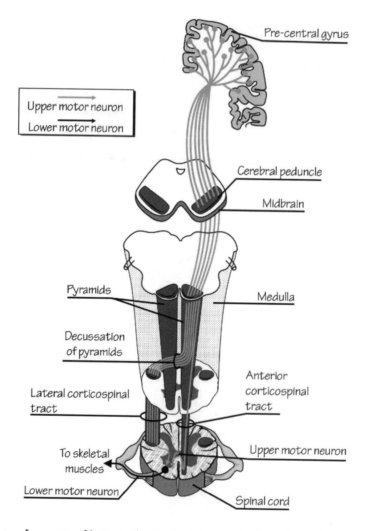

There are two pathways for motor fibers in the brain. They differ based on the kind of muscle control. You need different muscle control to move your arms and legs than you do to maintain your posture, so the nerve pathways for those functions are separate. For appendicular or extremity control, your brain uses the lateral corticospinal tract, with fibers that cross over at the pyramids in the medulla, before going down the ventral horn of the spinal cord to supply the muscle control over your arms and legs.

For control over your axial or postural muscles, the pathway is different. These fibers go down the anterior corticospinal tract, which has fibers that do not decussate. They descend further down the spinal cord in order to reach the level they need to be before exiting the CNS. Many of these fibers do decussate at the different spinal levels but, as mentioned, not all of them.

Why would this be necessary in the axial muscles but not the appendicular muscles? Think about it. Do you need to tell your right arm to relax before using your left hand to reach for a can of soda? Probably not. But does your brain need to tell your left-sided trunk muscles to relax as you bend toward the right? Of course, because you couldn't do the movement without that kind of coordination.

The Peripheral Nervous System

In this section, we will connect this central processing area of the nervous system with the peripheral nervous system, or PNS. The PNS is the working part of your nervous system. These are the areas that, in a corporation, need guidance from above but still go about their business, making sure the real work of the corporation happens on a daily basis. This part wouldn't function without the instructions from the higher-ups, but these same higher-ups in the corporation wouldn't get too far without the workers doing their job as well.

Before we get into what the PNS does and how it works, let's imagine what would happen if the two parts of the nervous system couldn't coordinate their activities. Actually, we don't need to imagine this, because there's a rare disorder where this happens. It's called *locked-in syndrome* and happens with a brainstem or upper spinal cord stroke. The CNS and PNS each work fine, but the two cannot communicate because the pathway of communication between them doesn't function.

These patients are essentially completely immobile except that their body has certain postures that reflect a lack of central nervous system input. The only muscles that work well at all are the eye muscles, because these work through the cranial nerve nuclei and don't involve the same brainstem pathways as the rest of the body. This is the only way we know their CNS is functioning. Through blinking movements, they can convey their thoughts and needs in the only way their physical bodies can allow them to.

The peripheral nervous system involves the motor and sensory parts of the body that are not directly in the brain or spinal cord. As soon as nerve fibers exit the spinal cord, they become part of the peripheral nervous system, which is actually made of fibers that are sensory or motor and that are related either to the somatic nervous system (the part you have the most awareness of) and the autonomic nervous system, the part you have limited awareness of.

Let's look at the spinal nerves, which are the closest part of the peripheral nervous system to the spinal cord itself. This is what one of these looks like:

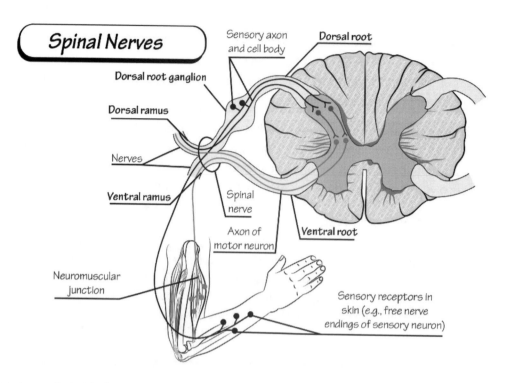

Spinal nerves are a hybrid of incoming, or afferent nerves, and outgoing, or efferent nerves. Nerves for the autonomic nervous system might also be incoming or outgoing and travel in these same pathways. You have 31 pairs of these spinal nerves coming off at nearly every level of the spinal cord. There are 8 pairs of cervical spinal nerves, 12 pairs of thoracic spinal nerves, 5 pairs of lumbar nerves, 5 pairs of lumbar nerves, and 1 pair of sacral nerves.

Since there are 8 cervical nerves but only 7 cervical vertebrae, the situation with these spinal nerves is different here than with the rest of the body. The first pair exits between the occipital bone of the skull and the first cervical vertebrae. Each subsequent one exits above the vertebra of the same number. At the C7 vertebra, there is a C7 pair above this vertebra and a C8 pair below. After this, the spinal nerve number exits below the vertebra of the same number.

As you can see, these do not leave the spinal cord at a single spot. Instead, there are two roots, one of which is the *ventral* or *anterior root* and the other the *dorsal* or *posterior root*. As we have talked about, the ventral root is mainly motor in nature, while the dorsal root is mainly sensory in nature. The dorsal root has a bump on it, which is the dorsal root ganglion where incoming sensory nerves connect. There isn't one of these on the ventral nerve root.

When these two roots come together, the bundle of fibers is called a *spinal nerve*, which is a mixed nerve fiber that leaves the spinal column through the holes shown earlier, the intervertebral foramina. Then this nerve separates again, this time into a *dorsal* and *ventral ramus*. The dorsal ramus travels to supply nerves to the back and areas around the spinal column itself. The ventral ramus has a longer pathway. These fibers go all the way to the ends of the arms and legs with multiple kinds of fibers.

There are tinier branches called *meningeal branches* that travel backwards through the foramina they just left in order to supply nerves to all of the ligaments, intervertebral discs, and other structures inside the spinal column itself. Another small branch, called the *ramus communicantes,* breaks off in order to have only autonomic nerves within it.

Finally, you should know that in some cases, nerve fibers from several levels combine outside the spinal column to make what's called a *nerve plexus*. These nerve plexuses are collections of interconnected fibers of all types and from different levels located in strategic areas along the spine. The goal of these plexuses is to provide a more complex system of nerves to supply key areas of the body. The four plexuses along the spinal column are the *cervical, brachial, lumbar* and *sacral plexuses*. This is what the brachial plexus looks like:

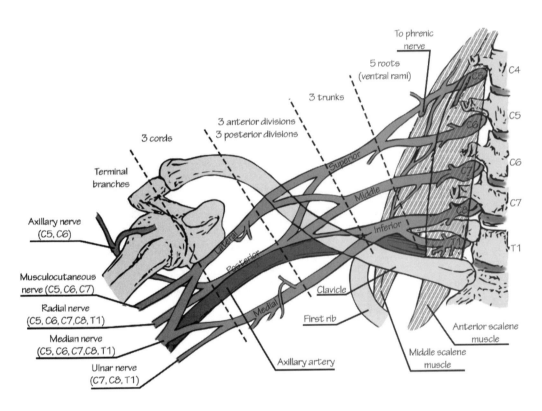

What Is a Spinal Reflex?

Spinal reflexes are interesting, because they involve the peripheral and central nervous systems together but generally do not actively involve the brain. Instead, these represent miniature relationships between sensory and motor nerve fibers that create an action you don't have voluntary control over. This is because the whole thing is a reflex arc that starts with a sensation in the periphery of the body, travels to the spinal cord, then leaves the spinal cord through a motor pathway to cause a movement. No part of this arc has to go up to the brain for processing in order for it to occur.

Let's do a typical somatic reflex arc scenario, which is what happens when a doctor taps your *infrapatellar tendon* just below your kneecap. When this test is done, the idea is to test the integrity of

225

the entire reflex arc involved in the knee-jerk reflex. The tap is the sensory input part of this. There are sensory receptors in the tendon that are attached to sensory nerve fibers. These go to the spinal cord and into the posterior horn of the cord.

Within the spinal cord itself, there are integrating centers that connect the incoming afferent fibers to outgoing efferent fibers. These synapse in the spinal cord, sending out an impulse to the lower motor neurons that carry out the final part of the reflex, which is to activate the muscles that jerk the knee. Of course, there are sensory fibers that do go to the brain, so you're aware that all of this has happened without the brain really participating in the reflex itself.

This is the basic pathway involved in the knee-jerk reflex:

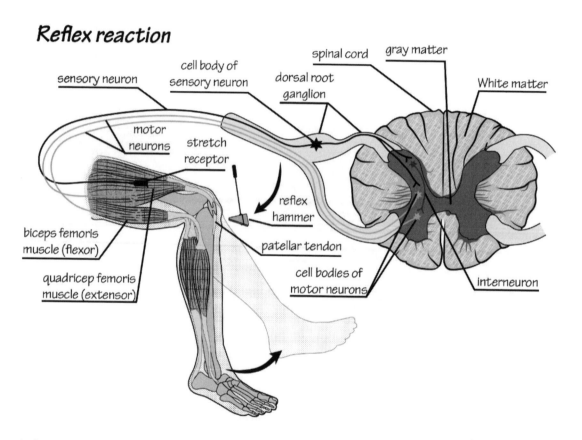

Cranial Nerves Are Peripheral Nerves

So far, we have talked about the spinal nerve component of the peripheral nervous system. These are the nerves that supply sensory and motor function to the body, largely the part below the neck. They're called *spinal nerves* because they're all linked to the spinal column in some way. This isn't true of the cranial nerves. Cranial nerves are still considered peripheral nerves, but they have nothing to do with the spinal cord and supply mostly the head and neck regions.

There are 12 cranial nerves that have nuclei in the cerebrum. Instead of going to the spinal cord, they exit at different points in the brain itself, leave the skull through their own openings or foramina, and supply the head and neck region in unique ways. Some are motor fibers and others are sensory fibers but the majority are mixed in some way, having regular motor, regular sensory, visceral motor (related to the autonomic nervous system), visceral sensory, and special sensory functions. The special sensory functions have their own category and involve things like vision, taste, smell and hearing.

Each cranial nerve is paired so that there's one on either side of the head and neck region. They're named by their main function and numbered I through XII using Roman numerals. The following image shows you where each one comes out of the brain. Their numbering system is such that the lower-numbered cranial nerves exit the brain more ventrally than the higher-numbered cranial nerves.

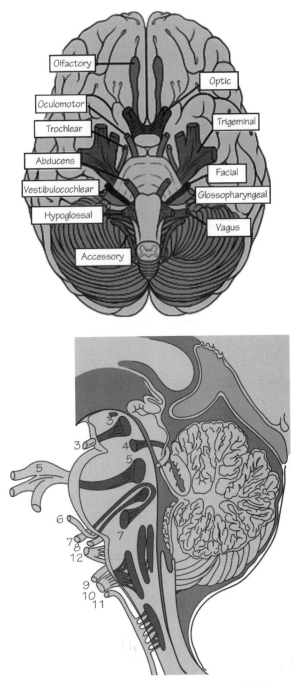

The names and basic functions of the different cranial nerves are as follows:

- **Olfactory nerve (I):** This is the nerve for the sense of smell. It is a special sense nerve only, where there are sensory receptors in the upper nasal passages that send signals to an olfactory bulb in the lower frontal nerve. The signal picked up is chemical, so that the nerve fibers respond differently to the various chemicals you smell. The fibers go throughout the brain directly and don't get processed by the thalamus first. These are the only sensory fibers that operate this way.

- **Optic nerve (II):** This is a special sensory nerve that governs your sense of vision. There are visual receptors in the eyes that travel back to the occipital lobe of the brain, where visual signals get processed.

- **Oculomotor nerve (III):** This nerve does several things. First, it operates multiple external ocular nerves that are involved with eye movements. It also operates the upper eyelid muscles that keep your eyelids open. One of its more unique functions is to participate in the *pupillary light reflex*, which is when your eyes constrict if a light shines in one of them. This is an autonomic nervous system reflex.

- **Trochlear nerve (IV):** This is a relatively minor cranial nerve that controls the movement of just one of the extraocular muscles. It's only a somatic motor nerve, so it has no sensory or autonomic functions.

- **Trigeminal nerve (V):** There are three large branches that supply all sensation to the face. The lower branch is called the *mandibular branch*. This is the only branch that also has motor activity, because it's the branch that allows for movement of the muscles of mastication, or your chewing muscles.

- **Abducens nerve (VI):** This nerve is also a minor cranial nerve that supplies just one muscle of the six total extraocular muscles: the lateral rectus muscle that moves your eye to the outside. It has a minor component in innervating the opposite medial rectus muscle in the other eye, so that your eyes move smoothly together and in coordination with one another.

- **Facial nerve (VII):** The facial nerve is also extremely important. It has multiple branches that provide motor sensation, or the ability to have voluntary movement over the facial muscles, particularly the muscles of facial expression. It also has autonomic nervous system activity to the *lacrimal gland* that controls your ability to have tears. There are also branches going to some of the salivary glands, so you can make saliva when it's necessary. Paralysis of this nerve or part of it will lead to droopiness of the face.

- **Vestibulocochlear nerve (VIII):** This nerve acts to control both hearing and your sense of balance. It is therefore a nerve related to the special senses. The cochlear component has receptors in the *cochlea*, or inner ear, that sends mechanical signals from the noises in the environment, turning them into electrochemical signals that are interpreted by the auditory cortex of the brain. There's a vestibular component that takes information about the position of the head in space when you're moving, to keep you balanced and prevent dizziness.

- **Glossopharyngeal nerve (IX):** This complex nerve affects the pharynx in multiple ways. It has sensory, motor and autonomic nervous system functions to this area. One branch goes to the *parotid gland* to control the secretion of saliva. There's a sensory branch that comes from the carotid sinuses and carotid body in order to regulate your cardiovascular functions. It provides sensation to the back of the tongue and the upper pharynx, and controls the special sense of taste for the back third of the tongue. If it's damaged, you might lose your gag reflex.

- **Vagus nerve (X):** This is a cranial nerve that has functions going far beyond the head and neck region. It is the major autonomic/parasympathetic nerve for the GI tract and helps with swallowing and speaking. It provides sensation to the external ear and *tympanic membrane*. It also aids in controlling cardiovascular functioning by innervating the *baroreceptors* that sense pressure change. There are branches that control the heart rate, adrenal gland function, pancreas, and parasympathetic peristaltic activity of the entire alimentary tract.

- **Accessory nerve (XI):** This is the main motor nerve for the movement of neck and shoulder muscles. It activates the intrinsic muscles of the larynx, so it plays a role in vocal cord function. It also has some sensory function related to these muscles.

- **Hypoglossal nerve (XII):** This is the main muscle that controls the motor activity of the intrinsic tongue muscles, or the muscles that move your tongue around. It's important in how well you are able to chew and swallow.

Fun Factoid: There are several interesting diseases of the cranial nerves you might have heard of. Two of these are called trigeminal neuralgia *and* Bell's palsy. *Trigeminal neuralgia is an inflammation of the fifth or trigeminal nerve—usually just a branch of it on one side. People with this disorder have severe, sharp pain in the face where the nerve travels. With Bell's palsy, the problem is with inflammation or dysfunction of the seventh cranial nerve or facial nerve. A person with Bell's palsy has no pain but will have a droop on one side of their face.*

Dividing up the Nervous System

We have so far talked about the nervous system in a relatively structural way, by dividing it into the central nervous system and the peripheral nervous system. In reality, these two systems work together to carry out many different functions of the body, from sensation to motor activity to controlling the muscles of the body.

We've basically been talking about the part of the nervous system called the *somatic nervous system*. The term *somatic* means "body." The basic parts of this functional unit of the nervous system include the somatosensory division, involving all the different physical sensations you might feel, along with the motor division, which is the part that decides voluntarily which muscles to move, and then proceeds to send signals to these muscles.

There are functional units of the nervous system whose parts receive contributions from the central and peripheral nervous systems acting together. The autonomic nervous system is largely involuntary, because the different components function outside of your awareness. It has two separate branches (the sympathetic and parasympathetic) that often oppose each other. Their role is to keep the internal organs regulated so that they can respond to changes in the environment.

These two systems have widespread activity to many end organs of the body. These include the pupillary responses, salivation, tearing, lungs or bronchial reactivity, heart rate, gastrointestinal motility, bladder function, sweating, blood vessel dilation, and sexual function. Here are the major activities of this system explained in pictures:

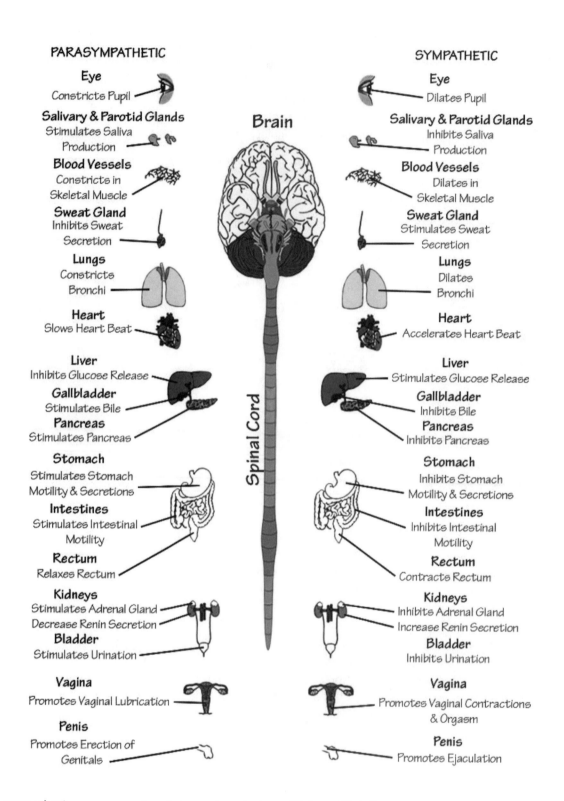

PARASYMPATHETIC

Eye
Constricts Pupil

Salivary & Parotid Glands
Stimulates Saliva
Production

Blood Vessels
Constricts in
Skeletal Muscle

Sweat Gland
Inhibits Sweat
Secretion

Lungs
Constricts
Bronchi

Heart
Slows Heart Beat

Liver
Inhibits Glucose Release

Gallbladder
Stimulates Bile

Pancreas
Stimulates Pancreas

Stomach
Stimulates Stomach
Motility & Secretions

Intestines
Stimulates Intestinal
Motility

Rectum
Relaxes Rectum

Kidneys
Stimulates Adrenal Gland
Decrease Renin Secretion

Bladder
Stimulates Urination

Vagina
Promotes Vaginal Lubrication

Penis
Promotes Erection of
Genitals

Brain

Spinal Cord

SYMPATHETIC

Eye
Dilates Pupil

Salivary & Parotid Glands
Inhibits Saliva
Production

Blood Vessels
Dilates in
Skeletal Muscle

Sweat Gland
Stimulates Sweat
Secretion

Lungs
Dilates
Bronchi

Heart
Accelerates Heart Beat

Liver
Stimulates Glucose Release

Gallbladder
Inhibits Bile

Pancreas
Inhibits Pancreas

Stomach
Inhibits Stomach
Motility & Secretions

Intestines
Inhibits Intestinal
Motility

Rectum
Contracts Rectum

Kidneys
Inhibits Adrenal Gland
Increase Renin Secretion

Bladder
Inhibits Urination

Vagina
Promotes Vaginal Contractions
& Orgasm

Penis
Promotes Ejaculation

The sympathetic nervous system is turned on during a "fight or flight" situation, in which great mental stress or physical danger is encountered. Neurotransmitters such as noradrenaline and adrenaline are released in response. An example of this would be if you come face-to-face with a vicious dog. Of course, you perceive the dog with your senses, but your response to the threat is entirely involuntary. The reaction you have is a visceral one in which your body prepares to run from the threat or fight it.

230

Typical things that happen include dilation of your pupils in order to see the threat more easily. Your mouth goes dry and your heart rate speeds up. There's a sudden shift in blood flow away from your digestive system and toward your peripheral muscles, so they'll have the energy to fight or flee. Your blood sugar goes up as sugar is mobilized. You begin to sweat. Your bronchial tree opens up and your bladder outflow is inhibited.

The parasympathetic nervous system uses the neurotransmitter acetylcholine as a mediator, to have the most effect on the body when there is no threat and your body is relaxed. This is the branch of the autonomic nervous system active in stimulating your digestive tract and its smooth muscles. An activated parasympathetic nervous system causes pupillary constriction, bronchial smooth muscle constriction, slowing of the heart rate, bladder emptying, and a general shift of blood to the digestive tract, where it aids digestion and promotes defecation.

Both systems must operate together and coordinate with one another when it comes to sexual responses. The parasympathetic nervous system helps with congestion of the sex organs in both males and females. The act of ejaculation in males is primarily regulated by the sympathetic nervous system. Orgasm itself isn't well understood, but it's probably some kind of autonomic sacral reflex.

Both the sympathetic and parasympathetic branches have their own ganglia. Besides these ganglia, there are nuclei in the brain itself, mostly in the brainstem area, that have input into the autonomic nervous system. The hypothalamus also coordinates with this system.

There are brainstem and sacral ganglia involved in the parasympathetic nervous system activity, located just outside the spinal cord. There's also a collection of sympathetic nervous system ganglia called the *sympathetic chain* that lies along the prevertebral tissues in the lower thoracic and upper lumbar regions.

The Enteric Nervous System

The enteric nervous system, or ENS, is the least understood part of the nervous system, yet it's very important to your daily functioning. This is the part of the nervous system related to the GI tract. The GI tract has a complex and vast nervous system, with its own neurotransmitters and the ability to interact with the rest of you. This system has input into your appetite, levels of thirst, and gastrointestinal motility. Through a complex interaction between bacteria, *neuropeptides* are released by parts of the gut, which can go on to affect your brain function and even your mood.

The GI tract can function by itself and has its own extensive network of nerves. Even so, there's a great deal of interaction between the GI tract and the central nervous system, so that one affects the other in complex ways. While the ENS is considered its own system, it is often considered to be a third branch of the autonomic nervous system that is influenced by each of the other branches.

There are literally thousands of ganglia within the walls of the entire alimentary tract, gallbladder and biliary tract. All these ganglia are interconnected and tapped into the local blood supply. About 200 to 600 million neurons are inside the GI tract, which is basically the same as the number seen in the spinal cord.

There are glial cells and nerve fiber bundles that project throughout the GI tract. The two major areas of enteric ganglia are the *myenteric ganglia* that are located between the muscle layers of the GI tract, and *submucosal ganglia* located inside the lumen of the GI tract. The myenteric ganglia form a plexus around the gut along its entire length. The intraluminal ganglia are mainly in the walls of the small and large bowel.

Both the sympathetic and parasympathetic nervous system are involved in enteric nervous system activity. The vagus nerve has a great influence on the entire GI tract, for example. The enteric nervous system also interacts with the prevertebral sympathetic ganglia. The endocrine system interacts with the GI tract as well—often on many levels. Finally, the GI tract supports immune function and is necessary for normal immunity to disease.

The net blood flow to the GI tract is regulated by the sympathetic nervous system, through its ability to constrict the blood flow to the alimentary tract. The idea is to give blood to the GI tract when it's needed for digestion and when a person is relaxed, while diverting blood away from the GI tract if you are stressed or in danger. This is why you don't digest food well when you're distressed or when your sympathetic nervous system is activated.

The ways the pancreas and stomach secrete things like gastric acid and pancreatic enzymes are all regulated by the enteric nervous system. These interact with GI hormones to control the way digestion happens. The GI tract is also important in the defense against toxins and pathogens, largely through reactions like vomiting and diarrhea, which are used to get rid of harmful substances.

Finally, the enteric nervous system interacts regularly with the central nervous system. There are incoming neurons from the GI tract that eventually become part of your awareness, including feelings of hunger, satiety, discomfort and pain. Things like nutrient and acid content in the GI tract, however, are not consciously understood by your brain but are regulated by the ENS. This is a two-way street so that the sight or smell of food, for example, will trigger your brain to act on the enteric nervous system activities, such as salivation and increased acidity in the stomach. While you can technically defecate without brain input, it takes this type of input to inhibit defecation until such time as it is both necessary and convenient.

The Act of Sensing with the Nervous System

While you are awake, you sense or perceive things all the time. You've probably heard of the five basic senses: touch, vision, hearing, smell and taste. These all require special sensory receptors throughout the body (or in specific organs of the body) that help you perceive your environment and make decisions about what to do with your perceptions—if anything. Your ability to create memories, and have a level of consciousness that means something, depends on what you perceive and how you perceive it.

You can divide your senses into two major categories, your somatic senses and your special senses. Somatic senses are found throughout the body—in your viscera, skin, soft tissue, muscles and joints. Some of these are called *mechanical receptors* or *mechanoreceptors*. They have the ability to detect various things you encounter all the time. Other receptors related to this part of the sensory system are called *nociceptors*, which detect pain, and *thermoreceptors* that detect temperature changes in your tissues. Some receptors are just free nerve endings that pick up sensations in the tissues, while others are special receptors structured for picking up certain things based on how they function.

Mechanoreceptors

There are four separate mechanoreceptor-type structures in the skin. They include the following:

- **Meissner corpuscles:** These are found in hairless skin and will detect pressure or dynamic touch. These have layers, or *lamellae*, made of Schwann cells. Inside the inner layers is the afferent, or incoming, nerve that fires when the skin is pressed on.
- **Pacinian corpuscles:** These can be found in the tissues beneath the skin, in the viscera, and in some connective tissue. They detect vibration and deep pressure applied to the tissues. These are structures that are layered like an onion, with lamellae that are surrounded by an outermost layer. Just inside this outermost layer is a fluid-filled space, with sensory nerve endings that respond to impulses you'd get from something vibrating the skin. These are rapidly adapting nerve fibers, so they can detect the speed of rapid vibrations.
- **Merkel's discs:** These are found near hair follicles and within the skin. They respond best to light touch and to the application of a steady pressure on the skin. These are found in the *epidermis* and line up along the dermal ridges of the fingertips. They are found mostly in very sensitive areas of the body, such as the genitalia, lips and fingertips. They help you detect the shapes of things applied to the skin and the edges of these shapes.
- **Ruffini corpuscles:** These are found within the skin and respond best when skin is stretched. They are elongated and encapsulated, located deep within the skin structures as well as in tendons and ligaments. They line up parallel to the stretch lines within the skin, so they are more sensitive to stretch than other receptors.

These different receptor types look like this:

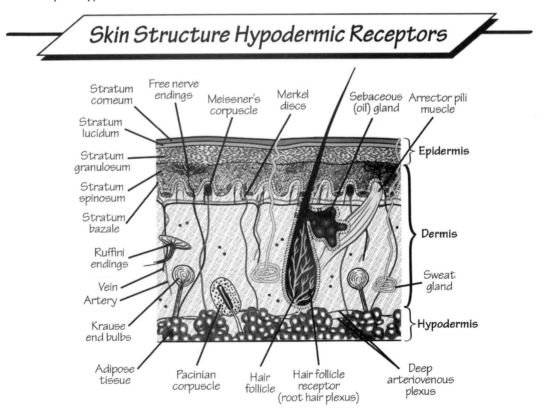

There are mechanoreceptors located elsewhere in the body, helping you to move your muscles and joints easily and often without having to look directly at the body part you are moving. These proprioceptors are located in muscles, tendons and ligaments. They tell you how you're moving and what your posture is. The three main proprioceptors you should know about are the *muscle spindles*, *Golgi tendon organs*, and *joint receptors*. Some of the input from these receptors mixes with neuron centers in the brain to keep you from being dizzy. This part of your brain is called your *vestibular system*, which is linked to your inner ear.

These are the major proprioceptors:

- **Muscle spindles:** These are tiny bundles of muscle fibers within each muscle that have a connective tissue capsule around them. There's a central multi-nucleated fiber called a *nuclear bag fiber* and smaller fibers called *nuclear chain fibers*. These special fibers have the ability to tell you how long each muscle is. You integrate the information from all your muscles in order to tell you what these muscles are doing without looking at them.
- **Golgi tendon organs:** These are receptors found within the tendons that attach muscle to bone. They respond to the level of tension in the tendon and also help decide how tense each muscle is within the body.
- **Joint receptors:** These are simple receptors within the joints that fire differently based on the tension within the joint. Like the other receptors, they help you determine if your joints are straight or bent and how much tension is on them.

Each of these different mechanoreceptors have specific characteristics that allow them to say a lot about the many perceptions your skin, subcutaneous tissues, and joints receive all the time. Some receptors are not very adaptive, meaning they fire rapidly and reset themselves so they can respond to the next incoming signal, such as those that detect vibration.

Other receptors are very adaptive, so they quit firing if the stimulus stays the same over time. It's how you're able to feel the sensation of your clothing on your body but don't notice it all the time. If the receptors for this type of sensation continually fired, you would perceive clothing all the time, which would be very distracting.

The benefit of having both these types of receptors is that you can receive information regarding both static and dynamic stimuli. Adaptive receptors respond quickly but briefly, while *tonic receptors* will fire as long as the stimulus continues to be present.

Your *Merkel discs*, Meissner corpuscles, Pacinian corpuscles, and Ruffini corpuscles are collectively called *high-sensitivity* or *low-threshold receptors*. These are extremely sensitive and are attached to fast-moving myelinated axons, so that the information they receive can get sent to your brain very quickly.

Nociceptors

Nociceptors are just free nerve endings without specialized structures that detect pain. Painful stimuli are detected by the nerve endings, resulting in a firing of a signal that is sent back to the brain for perception. There are cell bodies in either the *trigeminal ganglion* in the brainstem (for the detection of facial pain), or in the dorsal root ganglia on either side of the spinal cord (for pain sensation in the rest of the body). Nociceptor axons are only slightly myelinated or unmyelinated, so they act slowly compared to myelinated fibers.

Faster fibers are found mostly in the very sensitive parts of your skin. There are other nociceptors that are unmyelinated and respond best to chemical stimuli, certain mechanical stimuli, and temperature-related stimuli. The receptive fields of these nerve fibers cover a wide area of your skin, probably because it's more important to detect that pain is present than it is to say exactly where it's located.

There are actually two types of pain perception. One is called sharp pain, while the other is a delayed second pain that is duller. Fast nerve fibers can tell you quickly if something is sharp, while slow nerve fibers can tell you if the pain is dull.

When something very hot touches your skin, there are thermoreceptors in the skin that say, "Ouch, that hurts!" only when the temperature is too hot. Some do not detect pain but just detect the actual temperature. These are all free nerve ending receptors without any special receptors attached to the nerve endings.

The Special Senses

There are four special senses, which are smell, taste, hearing and sight. Each sense is associated with an organ that sends signals to the brain that you then perceive with special neuron areas located in the cerebral cortex. You can think of your eyes, ears, nasal mucosa and taste buds as being very specialized receptors that detect only certain things that your body and mind experience.

Your Visual Sense

The eye is the organ involved in your sense of sight or vision. Your eyes work by detecting light in various patterns, and then converting the light patterns they see into electrochemical impulses in the neurons of the retina. The retinal cells, called *rods* and *cones*, are *photoreceptors* that see photons of light energy by allowing the energy to change pigments within these cells. Depending on how much different frequencies of light activate these pigments, your entire retina will send off a blended signal that tells your brain what the image looks like, including its color.

Rods are super sensitive to light energy, so they get easily activated in low-light conditions. The downside is that they have only one pigment in each cell (called *rhodopsin*), meaning they aren't good at detecting color differences in low-light situations. In addition, rods are terrible at giving you a sharp image.

Cones have three different pigments in them, so they are good at detecting colored objects. They also have the ability to see images more sharply than rods. This is why there are many more cones near the *fovea* of the retina, where most of the light falls when you're directly looking at something. This is what your retina looks like:

Normal Retina

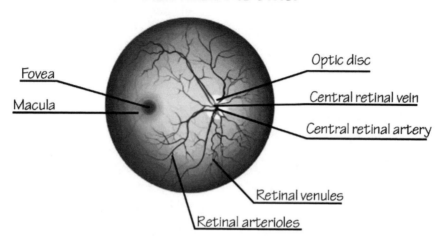

The part where of the retinal nerve fibers come together in order to send all of the information back to the brain is called the *optic disc*. The optic nerve extends from here back into the brain for information processing. The optic disc has no rods or cones. It's a functional blind spot in the retina of each eye. This is also called your *blind spot*.

The rest of the eye is all about getting the light and images into the eye in a clear way, so that everything you see isn't blurry. Light travels into the *globe*, or eyeball, starting at the *cornea*, which is a transparent covering over the *iris*, or colored part of the eye. Then it must go through a *liquid* or *aqueous humor* before going through the *pupil* and getting further sharpened by the *lens* of the eye. The lens is cool, because there are muscles attached to it that allow it to change shape, to see things at varying distances in your visual field. The *vitreous humor* is what fills the eyeball and gives it shape. It's more gelatinous than the aqueous humor.

This is what your eye looks like as light passes through the different layers.

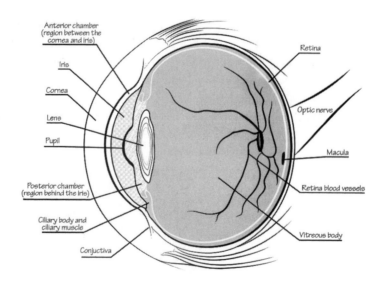

Your Auditory Sense

While your sense of vision is a chemical sense, because it involves photons that chemically change a pigment in the photoreceptor cells, your auditory sense is a special mechanical sense. It converts mechanical sound waves into electrochemical messages that go to the brain to tell you what it is you've heard.

It all starts at the *auricle*, or external ear. This is the part of the ear you see. It gathers sound and funnels it into the *external auditory canal*, which is the hole in your ear on each side of your head. The sound travels to the eardrum, or tympanic membrane, which vibrates just like a real drum does when you bang on it. This sends the impulse through the membrane and into the *middle ear*.

In the middle ear, you have a set of linked tiny bones called the *stirrup*, the *anvil* and the *hammer*, also called the *malleus*, *incus* and *stapes*. These take the impulse and focus it even further onto a smaller window called the *oval window*. This also vibrates, sending the sound impulse deep into the temporal bone where your inner ear is located. Your auditory tube, or Eustachian tube, will equalize the pressure within this middle ear cavity. This is what these structures look like:

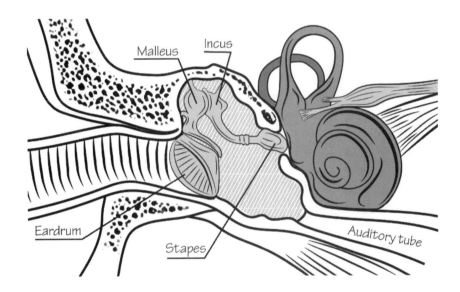

In the inner ear, you have two unique structures. There is the cochlea, which is shaped like a spiral seashell. Sound waves travel in the cochlea and bounce off hair cells that line this structure. Different frequencies of sound bounce off the hair cells in the different aspects of the cochlea. This breaks up many frequencies of sound into lower and higher frequencies that get processed separately. *Hair cells* are deflected or bent when the sound wave reaches them.

Certain bending patterns trigger these cells to depolarize, causing them to detect the sound and pass it on to the brain. Your brain puts all the hair cell information from the entire cochlea into one coherent message that you detect as words, bird sounds, or whatever it is you're hearing. You're able to hear low- and high-frequency sounds but only within a certain range. If you were a mouse, for example, you'd hear high-frequency sounds much better than low-frequency ones.

Your vestibular system is also a part of your inner ear, but it's sort of its own sense. It involves the *semicircular canals*, which also have hair cells. These cells act a bit differently. On top of the hair cells is a gel layer that has calcium-containing stones called *otoliths* on them. Depending on the direction your head is moving or on the way your head is tilted, gravity acting on these particles will deflect or bend the hair cells in certain ways. The overall pattern of bending throughout these semicircular canals will tell your brain if your head is moving and what position it's in. It looks like this:

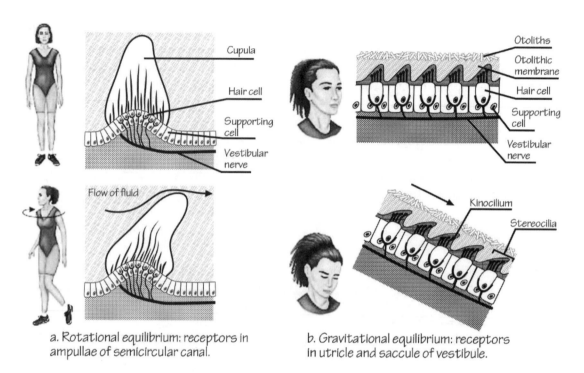

a. Rotational equilibrium: receptors in ampullae of semicircular canal.

b. Gravitational equilibrium: receptors in utricle and saccule of vestibule.

Your Sense of Taste

Your sense of taste is a chemical sense rather than a mechanical one. This is a complex sense in which you detect the chemicals dissolved in your saliva. You can detect the difference between something that is salty, sweet, bitter or sour. While these tastes can be distinctively perceived, there are many more possible taste perceptions, such as *astringency* with cranberries, fattiness, starchiness, pungency with hot peppers, monosodium glutamate, and metallic tastes.

While we eat bitter substances, this is not what this sense of taste is for. The sense of bitterness comes from things like *strychnine*, *quinine* and *atropine*, which are plant alkaloids that can be poisonous. Even though we eat bitter things, it is often an acquired taste that not everyone has. Evolutionarily speaking, the taste of bitterness was designed so you would avoid bitter things and their possible poisonous nature.

Your sense of taste is both qualitative and quantitative. This means that you can tell how much of a taste is present, as well as which taste is present. The sense of taste decreases with age, which is why many older people add more salt and spices to their food.

There is no truth to the myth that there are only certain areas of the tongue that sense the different types of taste. All the different taste sensations can be detected all over the tongue, but there are different regions of the tongue that have more or less sensitivity compared to other regions. It's true that sweet tastes are more easily detected on the tip of the tongue, and that bitter senses are more easily detected at the base.

Taste buds are found throughout your mouth and upper throat area. There are about 4,000 of them. Each taste bud itself has up to 100 taste cells, which are the main sensory receptor cells. The upper, or dorsal, surface of the tongue has 75 percent of all the taste buds in your mouth and papillae that help you taste better. This is what a taste bud looks like:

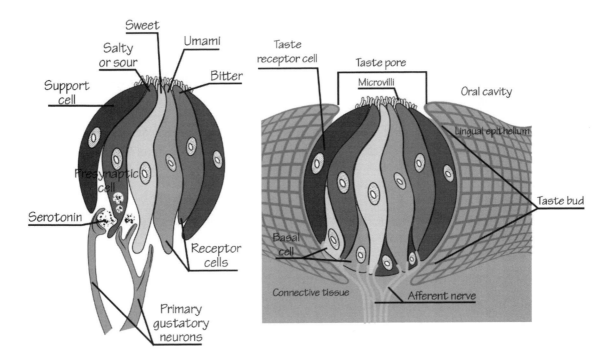

The taste buds are found on three different types of papillae. The taste cells within each taste bud will connect with any one of the three cranial nerves linked to taste, which are the facial, *glossopharyngeal* and vagus nerves. *Fungiform papillae* are all related to the facial nerve cells, while the *circumvallate papillae* are connected only to the glossopharyngeal nerve. The *esophageal* and *epiglottis-related taste buds* are linked to the vagus nerve.

The job of the brain in tasting is to integrate food concentration, identify the food, and establish the degree to which the food is pleasurable. Different cranial nerves respond in different ways to the various tastes. For example, the facial nerve responds the most to sugar and salt, while the glossopharyngeal nerve responds mostly to bitter and acidic substances.

As information is gathered from all of these places, you can integrate a lot of information about what a taste represents as a unique flavor. It would be impossible for one neuron by itself to make these decisions; so in order to assess the quality of food in the brain, you need to have input from all the taste-related cranial nerves, plus other information you perceive in order to make conscious decisions about what a type of food means to you.

Your Sense of Smell

Your sense of smell is called *olfaction*. It is evolutionarily probably your oldest sense. It acts differently from other senses because it isn't processed by the thalamus. It is a chemical sense like taste, because chemical properties of a substance trigger certain neurons in the roof of the nasal cavity to be activated.

The stimuli that trigger the sense of smell are called *odorants*. These interact with the olfactory epithelial cells inside the nose, sending signals from receptor cells in the nose up to the olfactory bulb in the frontal lobe. The neurons in the olfactory bulb send signals directly to the *pyriform cortex* within the temporal lobe, bypassing the thalamus entirely. Neurons in this tract go to the limbic system parts of the brain, such as the amygdala and hypothalamus. This is why smells elicit such emotional reactions.

Some odorant molecules are more easily detected than others. Lipids are more easily detected at low concentrations compared to water-soluble molecules. Attempts have been made in the past to characterize odors in ways that have been done for the sense of taste. The list of possible choices is much larger than with taste, so it is more complicated to study and partly explains why the sense of smell is so poorly understood in humans.

One other complication related to the study of the sense of smell is that many senses are perceived differently, depending on the concentration of particles in the air. We have different perceptions of rosewater in low concentrations depending on the amount of rosewater in other foods. In these situations, foods high in rosewater feel too "perfumy" to eat.

The sense of smell is basically the same in all people, but there those are of us who have a defect that leaves us unable to identify at least one type of smell. This is called *anosmia* and is rarely an ability to smell anything, but is usually just an inability to smell certain things. An example is the ability to detect cyanide, which is absent in 10 percent of people. This is a genetic defect that many people have.

When you smell things in the environment around you, there is the potential for the smell to cause other reactions in your body. You might salivate and increase your gastric motility when you smell good food, or gag and vomit if you feel like a smell is noxious in some way. Smell, while poorly understood, is a factor in regulating menstrual cycles in women, mother-and-child interactivity, infant feeding preferences, and sexual attraction.

Doctors believe we can smell about 10,000 different odorants. Your ability to smell things on the olfactory epithelium in your nose varies in different areas of this epithelium. Some things are detected in one area but not in another. When these neurons then project axons onto the olfactory bulb, they are also encoded spatially in what's called *space coding*. Space coding happens in vision and makes a lot of sense in how vision is interpreted, but it doesn't make as much sense in olfaction. There is also temporal coding in some species, in which smells are detected at different times, like a lingering odor after a stronger smell is detected.

The sense of smell begins at the olfactory epithelium in the nose. These olfactory neurons are bipolar with two axons: one goes to the nasal cavity while the other goes to the olfactory bulb. The tiny axon that goes out to the nasal cavity has a knob at the end that reaches out past the mucous layer of the nose. There are microvilli on this knob called *olfactory cilia* that pick up the odorant chemicals they are exposed to. This is what they look like:

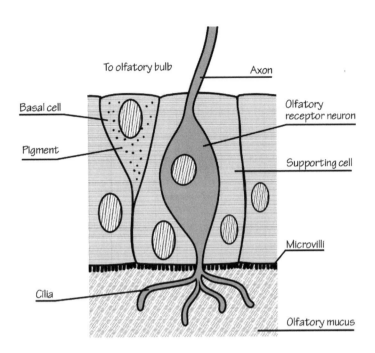

The longer axons from the olfactory epithelium must reach the olfactory bulb to start central nervous system processing of the smell. The olfactory receptor axons form bundles to make the *olfactory nerve*, which travels directly to the olfactory bulb without crossing over. The olfactory bulb is part of the CNS and can be found on the front and underside of the frontal lobe.

Inside the olfactory bulb, there are *mitral cells* that pick up the signal of the odor being detected. These cells are responsible for sending the information out of the bulb and into the rest of the central nervous system. These mitral cells send a dendrite into a *glomerulus*, or bundle in the olfactory bulb. This dendrite breaks into tufts that collect information from up to thousands of olfactory receptor neurons at once. This is what the whole process looks like:

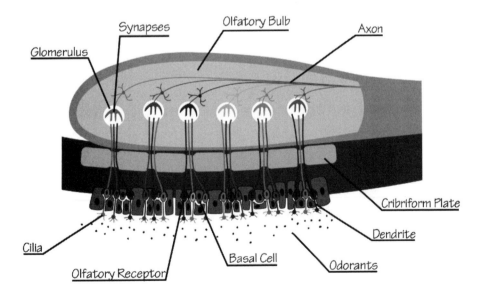

The mitral cell axons in the olfactory bulb form a small bundle of nerve fibers called the *lateral olfactory tract*. This travels to several areas in the brain, including the amygdala, the *entorhinal cortex,* the *olfactory tubercle*, and the accessory olfactory nuclei. Most of the axons, however, go to the temporal lobe in an area called the *pyriform cortex*. We don't know as much about how this pyriform cortex processes odors as we do about how other senses are processed.

To Sum Things Up

Your nervous system is perhaps the most complex of the major body systems you have. It can be divided into the higher-up regions of the central nervous system, or the CNS, and the lower working regions of the peripheral nervous system, or PNS. In addition, you can divide the nervous system into incoming fibers that relay sensory information to the brain, and outgoing fibers that send information to your muscles, viscera and glands.

Your autonomic nervous system regulates subconscious processes, such as those related to your heart, lungs, digestive tract, sweating abilities, bladder and sexual function. The two branches, the sympathetic and parasympathetic nervous system, often oppose each other and are activated under times of perceived danger (the sympathetic branch) and times of relaxation (the parasympathetic branch).

Your sensory system involves somatic senses from the skin, soft tissue, viscera and musculoskeletal organs, in order to detect things like pinprick touch, light touch, vibration, pressure, stretch, pain, and temperature variations. Your special senses are things like vision, hearing, taste and smell. All of these senses send information up to the brain, where there are multiple parts that collect the information, make decisions about it, and then send outgoing motor messages that affect how the body responds to these perceptions.

Chapter Twelve: The Endocrine System and Hormones

Sleeping. Eating. Gaining and losing weight. Sex and reproduction. These are all important things, right? They are exactly the things your endocrine system does for you. Your endocrine system is basically a system of glands, some of which are interconnected with each other in terms of what they do. Endocrine glands release hormones, which are chemicals similar to neurotransmitters except that instead of working across a small distance, they act over long distances.

In general, hormones are released directly into the bloodstream where they travel far and wide throughout the body. They are like emails that a company sends. Some are read by every customer the email reaches. Others go everywhere but are only read by a few interested people. Still others are sent only to special customers.

In the endocrine system, there is kind of an "unsubscribe function" like there is on certain email ads. If the customer is getting emails too frequently, they can send a message back to the advertiser saying they've had enough.

The endocrine system is more complicated than that, though. There is an opt-out system where the receiving gland or tissue can send its own stop message, but rather than completely stopping any further messages sent to it, the receiver can regulate the flow of subsequent signals. The whole thing is tightly regulated, so that the receiving gland or tissue gets just enough of the hormone it needs but never too much. As you will learn, this is called a *negative feedback loop*, which makes sure the sending glands allocate just enough of its hormones to the receiving glands, but never too much or too little.

The goal of the endocrine system is to regulate body functions. This is what it looks like when a hormone reaches its target or receiving cell:

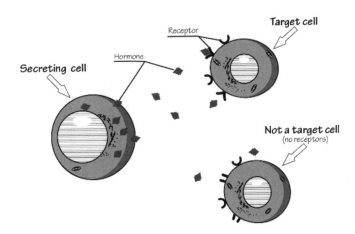

The target cell for each organ has specific receptors attached to their surface that will only bind to the hormone. The hormone attaches to the membrane in some cases, while there are other hormones that barge right in and do their action on the cell without the need for a receptor on the surface. Regardless of how it's done, the end result is that the hormone will change the activity or behavior of the target cell in some way.

In general, protein-based or other water-soluble hormones must have a protein receptor on the target cell for it to activate or change the target cell's behavior. The hormones attach to the receptor and cause a cascade of reactions to happen that affect the cell's activities. Changes in the number and sensitivity of receptors on the target cells can occur in response to low or high levels of hormones. This also helps to determine if and how much the target cell will be affected by the hormone. Lipid-like hormones, such as *steroid hormones*, can diffuse easily through the cell membrane, so their receptors are already within the cell. This is how a steroid hormone responds in the cell:

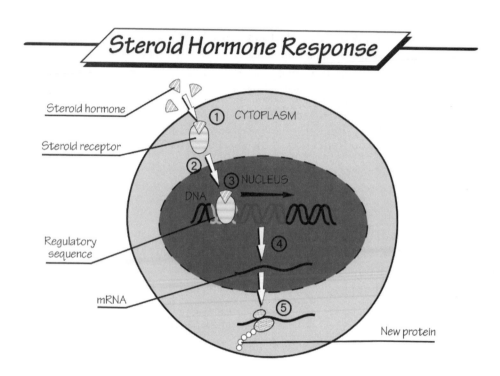

The blood levels of hormones consist of a balance between breakdown and excretion of the hormones by the kidneys, and secretion by the different endocrine glands. The kidneys and liver are the main organs that help to degrade hormones when they are no longer needed, breaking down the hormones into waste products and excreting them into the urine and feces. The half-life of the hormone and its duration of activity vary from hormone to hormone. Some hormones only last a few seconds, while others last much longer.

The Pituitary Gland and the Hypothalamus

The hypothalamus and *pituitary gland* work together in the endocrine system, even though the hypothalamus has other functions not related to the endocrine gland itself. You might have heard that the pituitary gland is the "master gland" of the endocrine system because it controls the activity of so many others. After you read this section, though, you might later decide that the hypothalamus is the master gland after all.

The pituitary gland is also called the *hypophysis.* It is an endocrine gland about the size of a single pea. It hangs near the hypothalamus at the base of the brain and sits inside a small bony cavity known as the *sella turcica*. It has two major parts: the *anterior pituitary gland* and the *posterior pituitary gland*. There

244

is also a minor third part, called the *intermediate lobe*. This is what it looks like, along with its relationship to the hypothalamus:

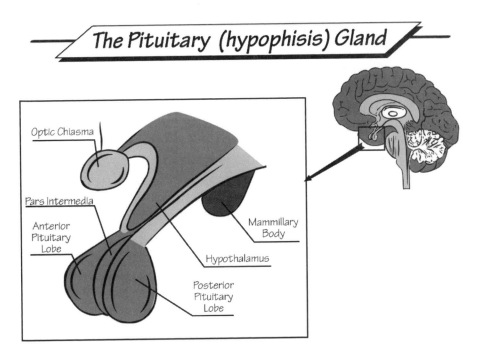

The posterior lobe is deeply connected to the hypothalamus by the *infundibular stalk*, or *pituitary stalk*. It actually makes none of its own hormones, but instead releases hormones already made by the hypothalamus, that are stored in the posterior pituitary gland prior to release. The two main hormones to remember here are:

- **Anti-diuretic hormone, or ADH:** This is also called *vasopressin*. Its job is to act on the kidneys in order to take up water when the body senses the concentration of the blood is too thick. It also constricts blood vessels to bring up your blood pressure. If this hormone isn't made or if the kidneys don't respond to it, you could develop a disorder called *diabetes insipidus*, in which you urinate a great deal, even if you aren't drinking very much.
- **Oxytocin:** This is also released from the hypothalamus and stored in the posterior pituitary gland. It is a complex hormone that is released in labor in order to force stronger uterine contractions during the birth process. It also causes the *let-down reflex* when a baby is nursing, and is sometimes called the "love hormone" because it seems to be released by mothers when they are near babies, or when you feel bonded to another person or animal.

Fun Factoid: One way to detect if a person has diabetes insipidus is to basically force a person not to drink anything. They are carefully watched to make sure they aren't cheating. If a person has diabetes insipidus for any reason, there will still be a lot of urine. This can quickly lead to dehydration, so it can be a dangerous test. There are two types of this disease. In central diabetes insipidus, you don't make enough vasopressin, while in nephrogenic diabetes insipidus, you make the hormone but your kidneys just don't respond to it.

The intermediate lobe of the pituitary gland makes just one known hormone: *melanocyte-stimulating hormone*. This is actually a group of hormones called *MSH*, *melanotropins*, or *intermedins*. These hormones can also be made by the skin. The trigger for the release of this hormone is UV light exposure, which tells the body to make more melanin. This is the skin pigment that darkens the skin so that you can be better protected from the damaging effects of the sun.

The anterior pituitary gland is the most complex part of the gland and is responsible for the regulation of several biological processes, including lactation, reproduction, growth, metabolism and stress. Since it makes so many hormones, you could easily call this the master gland, but the amount of some of the hormones it makes is actually controlled by the release of certain hypothalamic hormones. These are the anterior pituitary hormones:

- **Growth hormone:** This is the hormone that regulates growth, the making of proteins, and cell division. It also goes by the name *somatotropin*. GHRH (*growth hormone-releasing hormone*) and *GHIH* (*growth hormone-inhibiting hormone*, or *somatostatin*) are released in order to turn on or off the production of this hormone by the anterior pituitary gland. Too much of this hormone can lead to gigantism, or *acromegaly*, while too little of this hormone can cause dwarfism.
- **ACTH:** This is also called *adrenocorticotropin-releasing hormone*. Its production needs to be stimulated by the release of the hypothalamic hormone called CRH, or corticotropin releasing hormone. This goes on to stimulate the adrenal gland to release cortisol under stressful circumstances. We will talk more about this in a minute.
- **TSH:** This is also called *thyroid-stimulating hormone*, or *thyrotropin*. Its release is triggered by the hypothalamus, which releases TRH when thyroid hormones are necessary. The TSH directly stimulates the thyroid gland to release hormones that will stimulate the metabolism of every cell in the body.
- **FSH:** This is also called *follicle-stimulating hormone*. It is not released until the hypothalamus releases *gonadotropin-releasing hormone*, or *GnRH*. Its role is to stimulate the production of gametes or sex cells in both men and women, and to help these gametes reach maturity prior to fertilization.
- **LH:** This is called *luteinizing hormone* and also requires GnRH from the hypothalamus in order to be produced. When it reaches the gonads in males and females, it will turn on estrogen and progesterone production (in women) or testosterone production (in men).
- **Prolactin:** This is the hormone used to trigger milk production in women who are breastfeeding. It helps develop the milk-secreting glands after childbirth, and allows them to secrete milk when it is necessary during breastfeeding. It is inhibited by the dopamine neurotransmitter released by the hypothalamus and is triggered by the release of another hypothalamic hormone called *prolactin-releasing hormone*. It has no known purpose in men.

So, you can see why the pituitary gland is a major gland that secretes a great many hormones, but is also under the control of the hypothalamus in many cases. This is why it might be confusing to think of the pituitary gland as the master gland when the real master here is the hypothalamus.

The Thyroid Gland

Your thyroid gland is most responsible for the entirety of your cellular metabolism. That's because this small gland in your neck produces two hormones that act on nearly every cell in the body, controlling the metabolic activity of each. When you multiply this effect by all the cells in your body, the end result is that it affects your overall energy and metabolism levels.

The thyroid gland covers your upper trachea in your neck. It is called a *butterfly-shaped gland* because it has two lobes separated by an isthmus in the middle that connects the two lobes. The end result is that it looks like a butterfly. Its main job is to make two hormones called *thyroxine*, or *T4*, and *triiodothyronine*, or *T3*. Of these two, T3 is considered more potent and more likely to have a stronger effect on the cell. This is what the thyroid gland looks like in the neck:

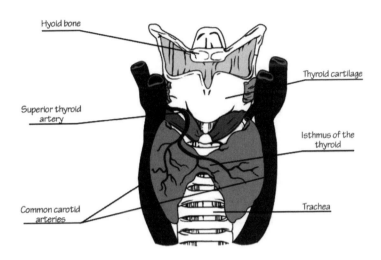

The best way to think about how the thyroid works is to look at what happens when it doesn't work right. If you think of metabolism as increasing the available energy and activity of a cell, and a lack of metabolism being a slowing down of the body's energy levels, it makes more sense what the thyroid gland does. The thyroid gland can be overactive or underactive, leading to either of these scenarios. Let's look at what an overactive and underactive thyroid gland looks like:

- **Overactive thyroid, or hyperthyroidism, effects:** bulging eyes (only seen in *Graves' disease*), goiter (swollen thyroid gland), nervousness, poor concentration, restlessness, increased appetite, irregular heartbeat, poor sleep, itching, fine or brittle hair, diarrhea, nausea or vomiting, hair loss, weakness, weight loss, heat sensitivity. It can look like this:

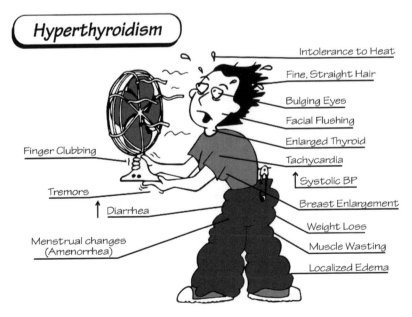

- **Underactive thyroid, or hypothyroidism, effects:** tiredness, cold insensitivity, constipation, dry skin and hair, weight gain, puffiness, hoarse voice, high cholesterol, muscle pain and weakness, irregular or heavy periods, thin hair, poor memory, goiter, slow heart rate, depression.

The purpose of the thyroid gland is to store, produce and release hormones into the bloodstream that are able to reach all the cells of the body. The thyroid gland uses iodine from the food you eat in order to make its two primary hormones.

Two glands in the brain—the hypothalamus and the pituitary gland—communicate with one another to maintain the T3 and T4 balance. The hypothalamus makes *thyrotropin-releasing hormone* (TRH), which signals the pituitary gland to release TSH. TSH is the message that stimulates the thyroid gland to make more or less of the T3 and T4. The entire process of making thyroid hormones requires iodine, which isn't a big problem near oceans (because fish and ocean water contain iodine), or where salt is routinely iodized.

The entire process is regulated by a negative feedback loop, which is a refinement of the idea of rejecting or unsubscribing to an email. In its simplest form, it looks like this: When the T3 and T4 levels are decreased in the bloodstream, the pituitary gland releases more TSH in order to tell the thyroid gland to make more thyroid hormones. If these thyroid hormones are too high, the pituitary gland releases less TSH to the thyroid gland in order to slow the production of these hormones.

In actuality, it is a little bit more complicated than that. High levels of thyroid hormones do not directly feed back to the pituitary gland but instead to the hypothalamus, telling it to reduce its TRH secretion. The pituitary gland is also blocked at the same time. This goes on to affect TSH levels, which then lowers the thyroid hormone levels. This is what it looks like:

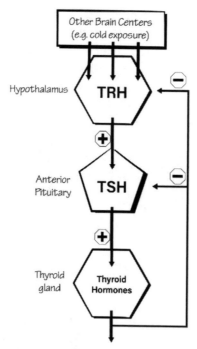

Target cells throughout body

Adrenal Glands

Your adrenal glands or *suprarenal glands* are somewhat complicated. There are two main sections, the inner *adrenal medulla* and the outer *adrenal cortex*. The medulla is important in the fight-or-flight response of the sympathetic nervous system because it releases epinephrine and norepinephrine under stressful circumstances. Your adrenal glands look like this:

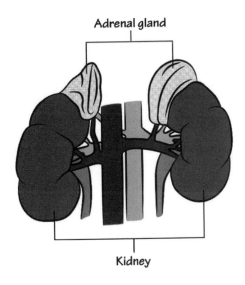

The adrenal cortex has some relationship to stress but in a different way. It has three layers that make the *mineralocorticoids* that regulate salt and water content in the body, certain reproductive hormones and hormone precursor cells, and cortisol, which is commonly released under stress. Here is a summary of these three cortical layers:

- **Zona glomerulosa:** This is where the mineralocorticoids like *aldosterone* are made. They regulate water, sodium and blood pressure levels in the body.
- **Zona fasciculata:** This is where the corticosteroids like cortisol are made, and where ACTH is most active in inducing the response. They influence blood sugar, blood pressure, and the immune responses to stress.
- **Zona reticularis:** This is where androgens are made that go on to make testosterone and other reproductive hormones. In women, a lot of the male hormones are made here rather than in the ovaries.

This is what these adrenal cortex layers look like:

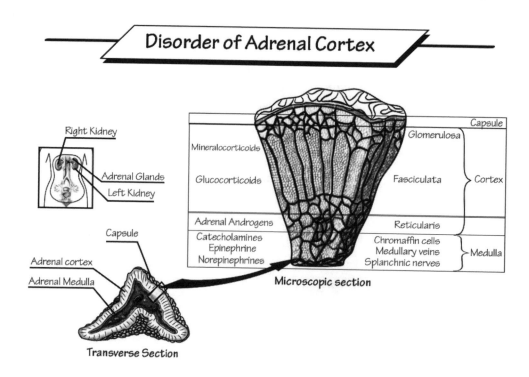

Disorder of Adrenal Cortex

Right Kidney

Adrenal Glands
Left Kidney

Capsule

Adrenal cortex
Adrenal Medulla

Transverse Section

Mineralocorticoids

Glucocorticoids

Adrenal Androgens

Catecholamines
Epinephrine
Norepinephrines

Capsule

Glomerulosa

Fasciculata — Cortex

Reticularis

Chromaffin cells
Medullary veins — Medulla
Splanchnic nerves

Microscopic section

The feedback system of the adrenal gland is called the *HPA axis,* or *hypothalamic-pituitary-adrenal axis.* It works in similar ways as the thyroid gland feedback loop. When cortisol is made in excess, it sends a signal back to both the pituitary gland and hypothalamus, causing CRH, vasopressin and ACTH levels to decrease. This self-regulates, so less cortisol is made.

Fun Factoid: The two main adrenal diseases you'll see are Addison disease *and* Cushing disease. *In Addison disease, you don't make enough cortisol, and in Cushing disease, you make too much of it. In true Addison disease your ACTH levels will be very high, because the pituitary gland is trying to nudge your adrenal glands as much as possible. One symptom is bronzed-colored skin; this is because ACTH is chemically similar to melanocyte-stimulating hormone, so your body will make more melanin, giving you a nice tan (even though in reality, this is a really serious disease that even President John Kennedy suffered from).*

In similar ways, if norepinephrine and epinephrine are high, this feeds back to the pituitary gland, causing a precursor cell that makes both ACTH and your pain-fighting endorphins increase. This turns on the adrenal cortex and lessens your sensation of pain when you might be injured in an attack, for example. This is a positive feedback loop system that only goes away when the threat disappears and you settle down.

The Pancreas

The pancreas has an endocrine component besides the exocrine part we already talked about in the section on the GI tract. The endocrine aspect is buried within the gland in clusters of cells called the *islets of Langerhans.* The function of this gland is not under the direct control of the pituitary gland.

While you've probably heard that the pancreas makes insulin, it actually makes additional hormones in different types of cells in these pancreatic islet clusters. This is what the islets of Langerhans look like:

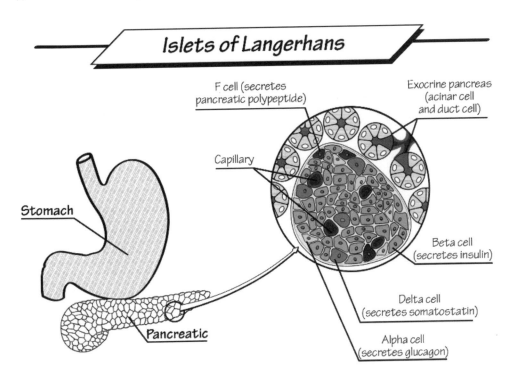

As you can see, the different cells of the Islets make different things. Each releases its hormones relatively independently of any of the others, although there are situations where one is inhibited as another is released. These are the endocrine hormones of the pancreas:

- **Insulin:** This is made by the *beta cells* of the pancreas. It is secreted as a pro-hormone called *preproinsulin* that needs to be processed before turning into insulin and being released to the bloodstream. When your blood sugar is high, these beta cells are triggered to make insulin. It acts on the cells of the body, including muscle, liver and fat cells, where it causes glucose to be taken up by the cell and used for metabolism. In fat cells, it promotes the conversion of glucose to fat. Its main job is to keep your blood sugar within as narrow a range of normal as possible.
- **Amylin, or diabetes-associated peptide:** This is made in the beta cells along with insulin when a person eats. It acts on the hypothalamus and the alpha cells in the pancreas in order to block *glucagon* secretion (which raises blood sugar). The communication between these two cell types delays both stomach emptying and sugar absorption. It also helps decrease your appetite by acting on the hypothalamus and its satiety center.
- **Glucagon:** This is made by the alpha cells of the pancreas. It directly opposes insulin and acts mainly on the liver cells. It breaks down glucagon to make glucose, and causes new glucose molecules to be made. As you can imagine, the result is that glucose levels rise. It gets released whenever the glucose level is too low in order to raise the blood sugar. It also breaks down fat.
- **Somatostatin:** This is made by the *delta cells* of the pancreas as well as the hypothalamus. You might remember that it acts to block growth hormone in the pituitary gland. It also blocks insulin secretion and the secretion of other hormones, such as gastrin in the stomach,

vasoactive intestinal peptide, glucagon, and TSH. It will act to block both insulin and glucagon at the same time, so its importance in blood sugar levels is limited.

- **Ghrelin:** This is made by the *epsilon cells* of the pancreas, as well as by the hypothalamus and the stomach cells. As you might remember, this is the hunger hormone that stimulates your appetite. It also increases growth hormone secretion in the pituitary gland. It will decrease insulin secretion in the nearby beta cells, so it has a complex relationship with your appetite and how you utilize the carbohydrates you eat.

- **Pancreatic polypeptide, or PP:** This is made by the *upsilon*, or *F, cells* of the pancreatic islets. It changes according to the dietary intake of nutrients you eat. The actual way this works isn't known.

As you can see, the endocrine pancreas has endocrine and paracrine functions. Paracrine activity of a cell means that it releases a hormone acting on a nearby cell rather than a faraway one. The main example of this is the fact that insulin and amylin both block the glucagon secretion in cells close to the releasing cell. Overall, while the function of the pancreas isn't 100 percent understood, the main goal is to regulate blood sugar levels.

The Pineal Gland

The *pineal gland* is in your brain and sometimes called the *pineal body*. It hangs near the thalamus in the midbrain area. Its main job is to produce melatonin in order to participate in your circadian rhythm. Your overall circadian rhythm is more than just about melatonin, because many of your hormones and your cellular metabolic processes vary regularly and predictably throughout the day.

The whole idea of circadian rhythm relates to the 24-hour day, even though your natural body circadian rhythm is probably greater than 24 hours (averaging about 25 hours in most humans). Melatonin helps to keep it closer to 24 hours per daily cycle because it is released as light levels drop, so that you can be awake during the day. The idea is that it promotes sleep, but it won't keep you asleep once you get there. People take melatonin when traveling in order to set their sleep pattern according to the time zone they are in, but usually this is what your pineal gland is supposed to do. Your pineal gland is shown in this image:

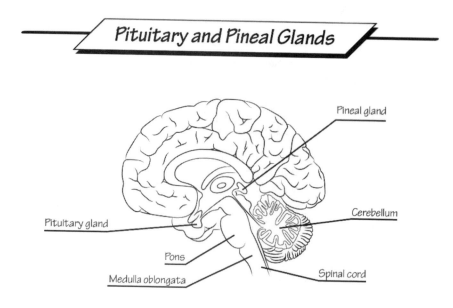

Pituitary and Pineal Glands

The Gonads

We will talk more about the gonads in the reproductive system chapter, but you should know that these are also endocrine glands. They are sometimes called the *sex glands* or *reproductive glands*. In women this gland is the ovary, while in men it's the testes. These are the main sources of gametes for reproduction in the body, but they also make hormones that regulate the menstrual cycle and control the expression of secondary sex characteristics, such as the breasts, hair growth patterns, and muscle mass in both men and women.

Your gonads are under the control of your anterior pituitary gland, which is itself controlled by the hypothalamus. The hypothalamus makes GnRH, or gonadotropin-releasing hormone, in both men and women, which turns on luteinizing hormone (LH) and follicle stimulating hormone (FSH)—also in both men and women. The entire process becomes different after that. Men respond differently to FSH and LH compared to women.

In women, FSH and LH are released in a cyclic fashion throughout the month. FSH levels rise to trigger maturation of early or primordial egg cells. As these mature, the estrogen level rises and the uterus prepares for a possible pregnancy. At the time of ovulation, the LH level surges to a very high level. This is the trigger for ovulation.

After ovulation, there is a remnant of the follicle where the egg was released during ovulation called the *corpus luteum*. It releases progesterone to support a pregnancy as well. If no pregnancy happens, the corpus luteum eventually gives out, the uterine lining sheds in the menstrual period, and the whole cycle happens all over again.

In men, there are testes where sperm cells are made and become mature. In the case of males, the process of GnRH release, FSH secretion, and LH secretion is the same. The difference happens in the testes. FSH triggers the precursor sperm cells, called *spermatogonia*, to divide through meiosis to make gametes or sperm cells. It also activates the *Sertoli cells* in the testes, which help in sperm cell maturation. LH is also necessary for this process because it encourages the *Leydig cells* in the testes to make testosterone to aid in *spermatogenesis*, or sperm formation. This is what spermatogenesis looks like:

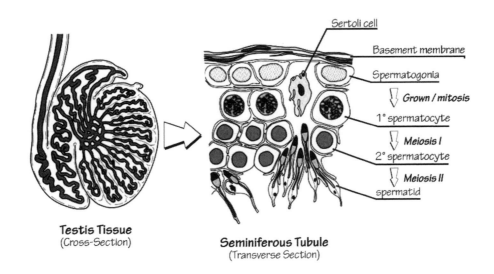

Testis Tissue
(Cross-Section)

Seminiferous Tubule
(Transverse Section)

There is a feedback loop that controls the amount of sex hormones and hypothalamic/pituitary hormones secreted by the body. In other words, the levels of testosterone, estrogen and progesterone will affect the hypothalamic and pituitary hormone levels. In some cases this is a negative feedback loop, while in others it is a positive feedback loop.

Parathyroid Glands

The *parathyroid glands* are four tiny glands embedded in the back side of the thyroid gland. Their only real job is to make *parathyroid hormone*, which is important in balancing your calcium levels. Calcium is necessary for bone growth and development. It's also important in muscle cell contraction; so if it's irregular, you will have muscle weakness, or *tetany*, which is spasm of the muscles. Because phosphorus is related to calcium levels, phosphorus levels in the body are regulated as well by these glands. This is what your parathyroid glands look like:

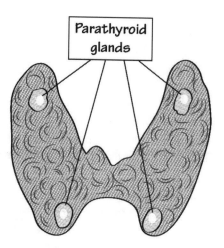

There is a counteracting hormone to parathyroid hormone called *calcitonin*. It's made by certain cells of the thyroid gland, but in real life calcitonin is not as important in regulating your calcium levels as parathyroid hormone.

Your parathyroid glands are always monitoring your calcium level. If it's low for any reason, the glands are triggered to release PTH, or parathyroid hormone. This hormone causes the bones to release excess calcium from its calcium stores in order to bring the level up. The metabolism of vitamin D is enhanced so that it allows more calcium to be taken up in the small intestines from your diet. Finally, the kidneys are triggered to make vitamin D *and* to enhance the reabsorption of calcium, so you don't lose as much in the urine.

If the parathyroid glands do not function well, the calcium level will be too low, and you could have muscle spasms, cramps, numbness around the mouth, tingling of your extremities, or even seizures. If the calcium level is too high, you can have osteoporosis because too much calcium is leaching from the bones. You might also have constipation, depression, an upset stomach, weakness and lethargy. Here's a brief summary of these symptoms:

	Hypocalcemia	Hypercalcemia
CNS	Irritability & anxiety Paresthesias Seizures Laryngospasm Bronchospasm	Decreased ability to concentrate Increased sleep requirement Depression Confusion and Coma Death
CVS	Heart failure	Arrhythmias Bradycardia
MSK	Muscle cramps	Muscle weakness

To Sum Things Up

Your endocrine system involves several different glands, all of which make hormones. Hormones generally get released into the bloodstream and will act on distant cells or tissues of the body. A few hormones are called *paracrine hormones* because they act on nearby cells rather than distant cells.

There are a lot of feedback loops in the endocrine system. Most feedback loops are called *negative feedback loops*. With these, it's a lot like telling an advertiser that you're getting too many emails about their product and they need to back off. With the endocrine system, though, it's more subtle than that, because the idea is to have just enough hormone rather than too little or too much.

There aren't as many positive feedback loops in the endocrine system, but one of them is the regulation of oxytocin in labor. When oxytocin starts the labor process, there is feedback that actually asks the posterior pituitary gland for more and more of the hormone, until the process stops when the baby is born.

Chapter Thirteen: The Immune System and Immunity

If you haven't yet figured out the anatomy and physiology mantra that "coordination is everything," maybe this chapter will convince you it's a real thing. Your immune system, just like the other systems we've talked about, is a highly coordinated system of organs, cells and biochemical processes that work together to keep you safe from pathogens. This is the system your body also uses to get rid of dead cells, damaged tissues, and even cancerous cells.

Your body needs protecting, just like a fortress or military base. In your immune system, you have barriers to keep out unwanted invaders. You have sentinels that sound the alarm and start fighting enemies as soon as one is detected. At the same time, the rest of the immune defense system is activated for more surgical strikes, where the enemy is more specifically targeted with less collateral damage involved. The goal is to protect your fortress and prevent any more pathogens from getting inside in the future.

The immune system can be divided into the *innate* and *adaptive immune systems*. Both of these work together in order to kill pathogens and destroy damaged cells. The innate immune system tends to work first. Think of this branch of the immune system as having the slogan "shoot first and ask questions later." A lot of this system doesn't care what the organism is, nor does it pay too much attention to preventing future infections. It is an attack system that goes after anything it doesn't recognize as belonging to the self, regardless of what it is. It works fast and mounts a strong immune response.

The second branch of the immune system is more like a sniper system. It is an intricate system of immune soldiers that knows exactly what it is that it doesn't want to be there, and can specifically kill that thing. This is the part of your immune system that makes antibodies to kill off specific pathogens, and has cells that can look at a specific cell, decide that it isn't healthy, and kill it in a special-ops strike that does little collateral damage to normal tissue.

The Organs and Tissues of the Immune System

Your immune system includes several primary immune system organs, various individual cells of the adaptive and innate immune systems, and some related organs called *secondary lymphatic tissues*. The major immune-related organs of your body include the bone marrow, the thymus, the spleen, tonsils, liver, skin, adenoids and lymph nodes.

The bone marrow is considered a main organ of the immune system because it makes all the main immune cells in your body. We talked about hematopoiesis before and about how white blood cells are made. These are some of the major immune-related cells of the body. They are either directly considered immune cells or will turn into immune cells at some point. Cells of the neutrophil line and lymphocyte line are extremely important in many immune system functions.

The thymus is important because it is necessary for the T lymphocytes in the immune system to become mature and functional. T cells arrive from the bone marrow for a thorough screening process before they can be let loose into the circulation to do their job. This is extremely important for several reasons.

First, you want any T cells you have to be healthy enough that your body recognizes them as being yours. The thymus has a process of determining if a T cell has the right receptors on it to belong to your own body. The next thing you want is for your T cells to also know that the rest of your body's cells belong to your body as well. The job of T cells is to recognize infected, damaged or cancerous cells in the body. If it accidentally thinks one of your healthy cells is foreign or damaged in some way, it will kill it, not realizing that the cell is normal.

This is exactly what happens in many autoimmune diseases. Your T cells accidentally think one of your normal cell types is abnormal, killing it and all the cells like it. The immune system tries to counteract this by having T cells select out and mature in the thymus. In fact, only a small percentage of T cells made by the bone marrow will survive this tough selection process.

The same process happens in the bone marrow and lymph nodes with the B lymphocytes, which are the antibody-producing cells. Only when the cell has passed through what's called *negative selection* and *positive selection* are they allowed to do their job.

With T cells, for example, positive selection answers the question, "Does this T cell belong in this body?" During negative selection, the question is, "Does this cell react a bit too strongly to cells it shouldn't?" If the answer is "no" to either question, the cell is destroyed and removed from the selection process.

The lymphatic system isn't exactly an organ, but instead a series of vessels that participate in the immune system. They often travel alongside arteries and veins, carrying pathogens toward lymph nodes or different types of immune cells from one area of the body to another. Lymph nodes are connected by lymph vessels.

Lymph nodes are found all over your body. You've probably heard of swollen glands in your neck, which are just swollen lymph nodes busy fighting a sore throat or other head or neck infection. There are lymph nodes in your stomach, armpits, thorax, groin and other places. They are like filters or traps, where pathogens get hung up along with a mixture of B cells, T cells, and other immune cells that cluster in lymph nodes in order to fight infection. This is a picture of a lymph node, where vessels run in and out as blood and lymph are being filtered:

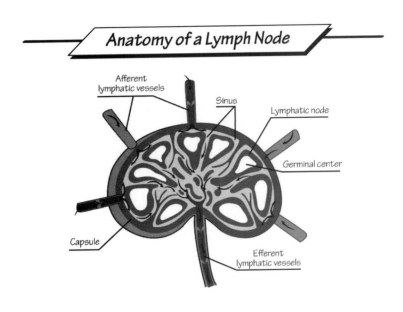

Anatomy of a Lymph Node

Afferent lymphatic vessels
Sinus
Lymphatic node
Germinal center
Capsule
Efferent lymphatic vessels

Your spleen is basically a large lymph node that does a great deal of blood filtering. There are antibodies made in the white pulp of the spleen so that certain bacteria collected there can be immediately removed. Old and dying white or red blood cells get filtered and removed by the spleen. In certain diseases where red blood cells are not shaped normally, the spleen doesn't see them as normal and destroys them, leading to anemia and a very large spleen.

While you can live without a spleen, having no spleen puts you at a greater risk for having certain bacterial infections, such as *encapsulated bacteria* that cause some kinds of pneumonia. This is the internal structure of the spleen:

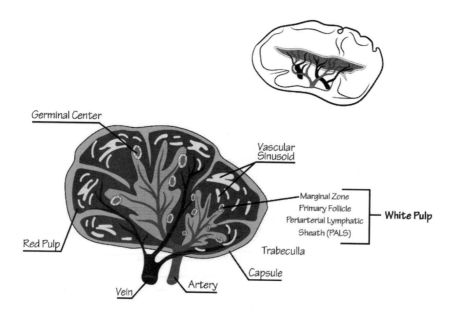

The tonsils and adenoids are also called the *nasopharyngeal tonsils* and the *palatine tonsils*, respectively. These are *secondary lymphoid organs*. The adenoids are in the nasopharynx and the palatine tonsils are in the oropharynx. Their job is to be sentinels where infections are most likely to get into the body, and act as the very first line of defense for pathogens you might inhale through your nose, mouth and throat. You can certainly live without them and their role is not completely understood, but they do have lymph node-like tissue where the destruction of pathogens can take place.

Fun Factoid: A little known area where lymphoid tissue like your tonsils is located is in the intestinal tract. This makes sense because you ingest a lot of potential pathogens all the time. In order to protect you, your intestines are dotted with Peyer's patches, which are small patches of lymphoid tissue that act as sentinels in this part of your body, in order to stop the invasion of pathogens before they can enter and cause a lot more damage to your system.

The liver has immune cells in it and makes some important proteins used in the immune system. It also filters out antigens that might come in when they get absorbed by the small intestines, so it acts similarly to other sentinels in the immune system to root out any unwanted foreigners in the body as soon as they enter.

Cells of the Immune System

There are a great number of cells involved in the immune response. Let's look at the different kinds involved:

- **Neutrophils:** Neutrophils are present in the circulation, and are able to leave the vessels during an active infection in order to fight infection in the tissues. These are some of the few cells that can *phagocytize*, or "eat," pathogens or other debris.
- **Natural killer cells:** These identify sick or infected cells, and destroy them through specific mechanisms that cause the damaged cells to die. Some of these are part of the innate immune system, while others are more direct and thus a part of the adaptive immune system.
- **Eosinophils:** These are the cells that participate most in the defense of parasitic infections or in allergies. They release substances that trigger typical allergic symptoms.
- **Basophils:** These are circulating cells that are active in certain types of inflammation. When activated, they release things like histamine and *heparin*. Histamine dilates blood vessels and make you itch or have hives, while heparin thins the blood to allow fluid and the movement of cells in and out of blood vessels during an inflammatory response.
- **Mast cells:** These are similar to basophils but are called *tissue-resident cells* because they are found in tissues but not in circulation. They are active in parasitic infections and release many of the same chemicals as basophils, which increase vascular permeability so that fluid and inflammation cells will leak out.
- **Dendritic cells:** These are also tissue-resident cells that do not circulate. They can develop from monocytes, among other cells. Their job is to eat pieces of pathogens, such as proteins called *antigens*, which are markers on the pathogens that will identify the pathogen as being bad or abnormal. Once found, they put these antigens on their cell surface, revealing them to phagocytic cells and other cells to direct their immune reactions. For this reason they're called *APCs*, or *antigen-presenting cells*.
- **Macrophages:** These come from monocytes, which circulate in the blood but ultimately leave the blood vessels, and turn into macrophages that eat pathogens and cellular debris. They also help in sending the message to other cells in the immune system so that they'll engage in fighting pathogens.
- **T lymphocytes of the CD8+ type:** These are types of T cells called *cytotoxic T lymphocytes*. These are the activated T cells that kill cells that are damaged, cancerous or infected.
- **T lymphocytes of the CD4+ type:** These are also called *T helper cells*. They coordinate your immune response against bacteria by secreting substances that will help to enhance the immune response. Different types will secrete their own type of immune-related molecules.
- **T reg cells:** These are a subtype of helper T cell that actually modulate or temper the immune response so it doesn't get overactive. They help to maintain your tolerance to your own immune system so you don't fight against your healthy cells.
- **B cells, or plasma cells:** These are cells of the adaptive immune system. Their main job is to make specific antibodies that help in different ways to neutralize or destroy pathogens. They also present antigens to T cells so that T cells can do their cell-killing job. In that sense, these are also antigen presenting cells, or APCs.
- **Memory cells:** These can be B cells, T cells or NK cells. They will circulate for a long time in the body, and have ways of remembering that an infection by a certain pathogen has occurred. The goal of these cells is to make sure that, if the same pathogen is encountered again, your body

will recognize it and mount a large response to fight it. This way you might never feel sick, or might have a milder case of the infection in the future.

How Your Immune System Functions

The best way to understand how the immune system functions is to see what happens if a pathogen gets into the body. From this scenario, you can see the different parts of the immune system that interact with one another in order to fight the pathogen—starting from its first entry into the body and ending with memory cells that help to prevent a similar reaction from occurring in the future.

The Barrier System

The first step in fighting an infection involves your *barrier system*. This would be like the walls of your fortress or the moat you build around it. The whole goal of this part of your innate immune system is to create an environment where bacteria, viruses and other substances cannot even gain entry.

One obvious barrier system is your skin. The skin consists of a thick keratinized epithelial layer that keeps most pathogens from getting through to your deeper tissues. The problem starts when you get a break in the skin somehow—through a cut or scrape, for example. This becomes a portal of entry and is where you'll often see redness and warmth as your body fights the infectious organisms that are getting through the broken skin.

Fun Factoid: Your whole inflammatory response comes as part of the innate immune system. When you think of inflammation, you probably think about pain, redness, warmth, swelling, and loss of the area's function. Each of these things is a nonspecific response, involving parts of the immune system that go in to fight the infection as soon as it's recognized.

Other barriers you have that prevent organisms from getting in and setting up shop include the epithelial cells that line the bladder as well as the respiratory, GI and reproductive tracts. In some cases, as in the respiratory tract, there are cilia along with a mucus layer that keep pathogens out, partly by pushing bacteria and debris up and out of the respiratory tract. There are special antibodies within these tissues (called *IgA antibodies*) that are particularly capable of inactivating pathogens at the source where they're most likely to enter.

Your tears, saliva and urine are also part of the barrier system. Tears wash away pathogens all the time from the surface of your eyes, and urine flushes out your bladder in order to get rid of pathogens in this organ. Saliva has antimicrobial substances in it that prevent infections, and a woman's vaginal pH is set so that most dangerous pathogens do not thrive well inside it.

The Inflammatory Response

One of the first things that happens in an infection or possible infection is what's called the *acute inflammatory response*. If you've ever had an infected skin wound, you know what most of the symptoms of this response look like. Any chemical, physical or microbial injury to your tissues can set up this response, which looks a lot like this:

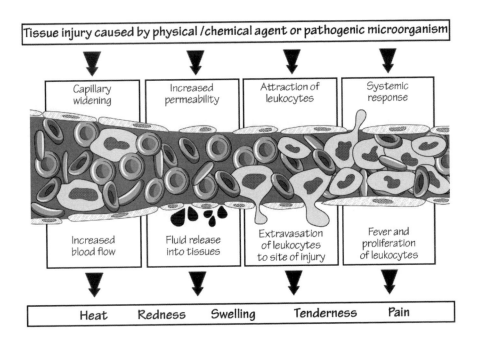

Acute inflammation of a tissue (like your skin) is pretty easy to spot. There are five main symptoms you will experience when a break in your skin has led to pathogenic invasion by microorganisms:

- **Pain:** This can be something you easily notice, or something you feel only when you touch the specific area that is infected.
- **Redness:** This happens because the blood supply to the area increases, leading to more redness.
- **Warmth:** This also comes from an increased blood supply, and from *pyrogens*. This may mean you have a fever from the infection.
- **Loss of function:** This means that you will not be able to move the affected area as much or, if the area is your respiratory tract, you might lose your sense of smell or have a hard time breathing.
- **Swelling:** This happens when the capillaries in the affected area become permeable to fluid, which floods the tissues, causing them to swell.

Notice that they happen automatically and are nonspecific. This means that the reaction happens regardless of what the pathogen is, and that it's part of your innate immune system. This is the "hit-and-run" part of your system that doesn't care what you're infected with.

The first reaction is that your blood vessels constrict, which is more to stop the bleeding than anything. Then they dilate in order to increase the blood flow to the area. The blood vessels, especially the capillaries, increase in permeability so that fluid, electrolytes and infection-fighting cells can leach out into the tissues. Clotting factors will leach out too, in order to keep the pathogens walled off into a specific area.

White blood cells, especially neutrophils, become anxious to fight off the pathogen. They escape the blood vessels, travel into the tissues, and begin eating whatever they can that isn't normal tissue. They will eat pathogens and any damaged tissues or cells. There are substances released by the tissues and by neutrophils called *chemotactic factors*. These are messenger molecules that say, "Help! Bring on more soldiers to fight this thing!" This draws in large numbers of infection-fighting cells to the affected area, so that all hands are on deck to fight the invasion.

Other chemicals that are a part of this response include histamine, which causes blood vessels to dilate and increases the blood vessel permeability. Mast cells in the tissues and basophils in the bloodstream will release histamine as part of the inflammatory response. There are other chemicals called *cytokines* that participate in different ways in the inflammatory response. These are released by several different types of cells in damaged tissues.

Prostaglandins are chemicals produced when tissues are injured. When there are prostaglandins in the tissues, you will have a sensation of pain. Prostaglandins also cause fever, localized heat and increased vascular permeability. The nonsteroidal drugs you take for inflammation, like aspirin and ibuprofen, work by being anti-prostaglandin drugs.

Inflammation happens very quickly—within a few hours or days of a break in the skin or mucous membranes. It can spread to involve symptoms throughout your whole body, but generally passes once the rest of your immune system gets on board and the damaged tissues are cleared out.

Innate Immune System Cells

In the evolution of the immune system in animals, the innate immune system is older than the adaptive immune system. Innate immune system cells need to know just one thing: Is this thing in the body something belonging to the self, or is it foreign in nature? It turns out that each cell of your body has markers called *MHC*, or *major histocompatibility complex markers*. There are two classes of these MHC markers, but each of them is unique to your body. If any cell, human or otherwise, does not have matching MHC markers on it, it's like not having the password to a secret club. In the case of your body, it reacts negatively to this and fights back.

Fun Factoid: These MHC markers are super important to things like blood transfusions and organ transplants. A transplanted organ must have as much of the same MHC markers as the person receiving the transplant for the best chance of a successful organ transplant. This explains why these procedures are best done when siblings donate organs to each other. Your MHC markers are inherited, so there is a good chance that one of your siblings is a good match for you, or at least a reasonable partial match. If they do not match, rejection or a reaction can occur.

Most cells of the innate immune system (including dendritic cells, macrophages, natural killer cells and neutrophils) have certain receptors on their cell surface called *Toll-like receptors* (*TLRs*). These belong to a family of receptors whose job it is to say if a certain pathogen is dangerous or not. They bind to interesting proteins found on many types of pathogens called *PAMPs*, or *pathogen-associated molecular patterns*. If a PAMP protein binds to a Toll-like receptor on an immune cell, the immune cell sees "I am something dangerous" on the pathogen that has the PAMP and starts the immune response.

A whole lot of things happen when a foreigner is detected. Cytokines that signal other cells to jump in and fight the infection are released by white blood cells, causing some immune cells to become more effective at killing pathogens. There are cytokines called *interferons* that block the replication of viruses, and still other cytokines that regulate the immune response so it isn't as hyperactive. *Chemokines*, or chemotactic factors, are also types of cytokines.

The major mechanism the innate immune system has in the defense against pathogens is called *phagocytosis*, or *the cell-eating process*. These are cells that are trained to eat and digest anything that isn't recognized as being part of the self. This is what phagocytosis looks like:

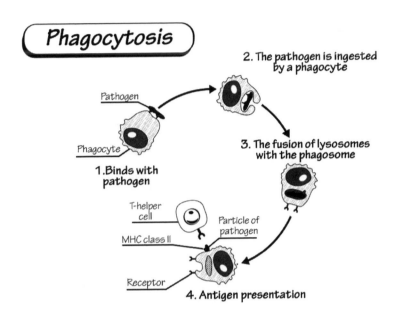

One thing you'll notice on this picture is that the pathogen isn't just eaten and forgotten about. If the phagocytic cell is also an antigen-presenting cell, the antigens or proteins on the surface of the pathogen are spit back out onto the surface of the APC.

The foreign antigen doesn't sit there by itself but is matched with an *MHC class II molecule* on the phagocytic cell. This is so that a T helper cell can recognize both the fact that the phagocytic cell is one of its own self cells *and* that it is attached to a foreign antigen that needs further addressing—this time by the adaptive immune system. This is one place where the innate immune system and the adaptive immune system work together.

That's what makes APC cells so important. The cells of the innate immune system that have the ability to participate in phagocytosis include macrophages (which come from monocytes originally), neutrophils, eosinophils and dendritic cells. Not all of these are APC cells, though. Only macrophages, dendritic cells and B cells are considered APC cells. The dendritic cells and macrophages are both cell types that will bridge the gap between the innate immune system and the adaptive immune system.

Natural killer cells of the innate immune system don't have the ability to kill pathogens by themselves. Instead they identify any host cell that has been infected, particularly by a virus, so that the entire infected cell and the pathogens in it are killed. They do this by being able to see that the infected cell's MHC proteins are no longer normal.

The complement system is part of the immune system as a whole. This means it can be active in the innate or adaptive immune systems. It involves numerous proteins that operate on each other in a cascade in order to have different immune responses.

The *complement cascade* is messy, because it can be triggered by more than one thing and involves a lot of different proteins acting on each other. If there is an infection, this cascade gets activated in order to help fight the infection. The goal of the complement system is to create a *membrane attack complex*, or *MAC*, on the microbial organism to target it for destruction. This image gives you a rough idea of what it can look like:

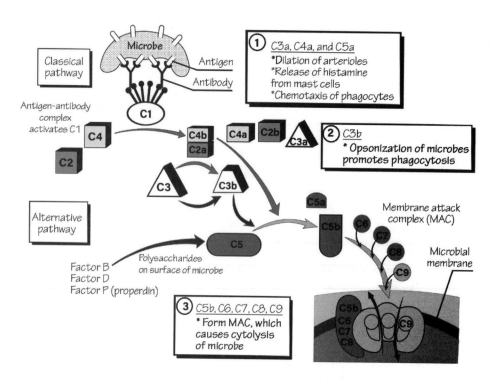

There are more than thirty different proteins in the complement system that all act in a certain sequence. They can help your immune system directly kill bacteria, make it easier for phagocytes to eat them, neutralize viruses, attract phagocytic cells, promote the formation of antibodies, and form complexes of dead cells and antibodies that are easy targets for destruction by immune cells.

The Adaptive Immune System

Your innate immune system is a pretty good system in the initial stages of an infection, but it isn't good at making direct hits at specific targets and remembering whether you've had an infection before. The adaptive immune system mainly involves T cells and B cells, both of which are types of lymphocytes that have specific responses to an infection.

The B cell part of the adaptive immune system is called the *humoral immune system* because it involves molecules more than it involves cells. B cells start out naïve, meaning they have never been exposed to any type of pathogen. In the lymph nodes or in your circulation, they can encounter an antigen, usually one that is presented to them as part of the innate immune system and the antigen-presenting cells.

Once activated, the naïve B cell becomes a plasma cell. This means it is an active working cell in the immune system.

B cells have certain receptors on their surface, called *B cell receptors* or *BCRs*. These can bind to antigens or foreign proteins and can detect that they're abnormal. The BCRs bind to the antigen, take it up into the cell, and process it so that it can either signal other cells to react to the antigen or make antibodies to fight off the organism where the antigen originated. This is what a B cell receptor looks like:

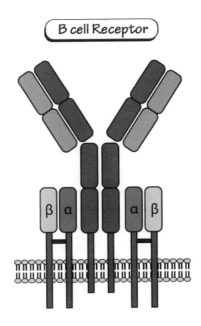

Plasma cells make different kinds of antibodies. Some stay on the surface of the B cell, while others are released into the circulation. The different types of antibodies are called *immunoglobulins*. The immunoglobulin types are *IgM*, IgA, *IgD*, *IgG* and *IgE*. These have various immune functions. This is what these different antibodies look like:

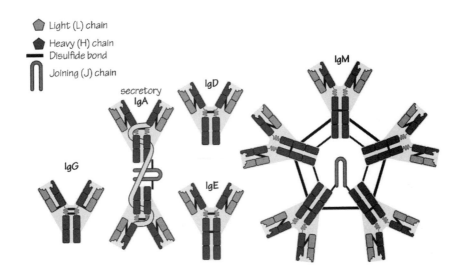

IgM antibodies are the first to be made in the immune response. Their job is to form large complexes that are more easily destroyed by the immune system. IgG antibodies are the most common type in the blood. They come on after IgM antibodies are made and have the ability to cross over the placenta, to help in preventing infections in a fetus. IgA antibodies are secreted onto mucosal surfaces to provide surface responses to antigens and pathogens. IgD antibodies stay attached to the B cell, and IgE antibodies are most active in allergic reactions or if the pathogen is a parasite.

Antibodies tend to be very specific in the way they attach to a certain pathogen. Once the antibodies are made and released by B cells, they bind to the pathogen at the antigenic sites just like a lock-and-key mechanism, so that only those things the antibody binds to get acted on by the immune system. Pathogens with antigens that have antibodies attached to them are usually doomed to be killed by the immune cells in a very specific way.

T cells are also important in the adaptive immune response, during a part called *cellular immunity*. This is because it acts from cell to cell in order to kill non-self cells, or those that are infected, damaged, sick or cancerous. T cells are always trolling for these kinds of cells so that they can kill them.

Fun Factoid: How many cancerous cells are created in your body every day? Experts think there are thousands made on any given day. Because your T cells are always patrolling the body for these cells, you don't get cancer nearly as much as you otherwise would if you didn't have this important function. Many new cancer treatments are directed not so much at destroying cancer cells but at improving your immune system's activity, so that your cells can selectively kill the cancer cells directly.

Remember that the body knows the difference between a self cell and a non-self cell because of the MHC molecules on the cell surface. If the MHC molecule on any cell has the secret code, it's left alone by the T cell. If the MHC molecule is a non-self molecule, or if something has happened to the cell that changes how the MHC looks to the T cell, it will destroy it because it sees it as dangerous.

This is how many virally infected cells get killed by the immune system. If a cell is cancerous (which happens a lot more than you'd think), the cell is marked as abnormal and destroyed as well. The cytotoxic T cell, or CD8+ cell, is the cell type that does the actual killing part, while *Treg cells* and CD4+, or helper T, cells assist in the process to make sure it goes smoothly.

So, how does a cytotoxic T cell kill an injured or infected cell in your body? It has several mechanisms to do this. One way is to release enzymes called *granzymes*, which are enzymes that are toxic to cells, causing cell death. The second way is to make a protein called *perforin*, which forms a hole in the target cell, allowing the granzymes to enter and cytoplasm to leak out. A cell that dies by this method is said to undergo *apoptosis*. The entire process of T cell destruction of a damaged cell looks a lot like this:

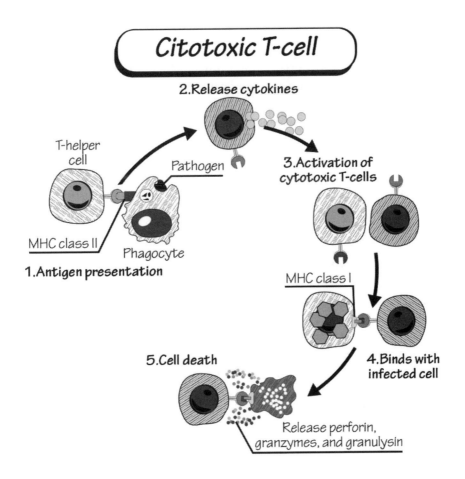

Citotoxic T-cell

2. Release cytokines

T-helper cell

Pathogen

3. Activation of cytotoxic T-cells

MHC class II

Phagocyte

1. Antigen presentation

MHC class I

5. Cell death

4. Binds with infected cell

Release perforin, granzymes, and granulysin

Both T cells and B cells have the ability to have immunological memory. These are called *memory cells*. When you've had an infection and your body has fought it off, antibody-making B cells leave behind some of their kind as memory cells that will remember that the infection has occurred. If you come in contact with the same infection again, those memory cells become active plasma cells to make antibodies to fight the infection without delay so that you don't actually get sick the second time around. T cells also have some of their own kind left as memory cells for long-lasting immunity. If you get a vaccination, the antigens are provided for you in the shot without actually having the infection, so you'll make preventative antibodies in an active way without ever having the disease.

When you get an infection or have a vaccination, you develop what's called *active immunity*. It's called *natural immunity* if you become immune because you actually got the disease. When a baby is born with antibodies that have crossed the placenta from the womb, that baby will have *passive immunity* and will be relatively immune to the same diseases as the mother. This is extended with breastfeeding because antibodies are passed from mom to baby as well. This type of immunity is temporary; the antibodies disappear within six months of a baby's birth or after stopping breastfeeding.

To Sum Things Up

Your immune system is a coordinated effort your body puts forth in order to keep you healthy. It is the bodily system responsible for fighting infections of all kinds, getting rid of damaged tissue, and even fighting cancer in your body. There are several parts to the immune system that work together to protect you from the potentially hostile environment around you.

Immunity starts with a barrier system, that attempts to keep infection out physically and chemically before an infection can develop in the body. The next step is to allow the innate immune system cells to take over, by destroying any pathogens not identified as self cells or that are viruses. This is a nonspecific system of immunity that will "shoot first and ask questions later," killing indiscriminately if the pathogen is felt to be dangerous in any way.

The adaptive immune system is more specific to a certain pathogen. It involves mainly B cells and T cells that target only those pathogens that have been labeled as having foreign antigens. B cells make antibodies that help promote the immune response to a pathogen, while T cells participate in cell-mediated immunity and will destroy targeted cells using destructive enzymes.

Active immunity means you are immune to an infection because you've had the infection already, or because you're immunized against a certain disease. If you receive immunoglobulin treatments in a hospital setting, or if a baby gets born with antibodies against the different diseases the mother has had before, this is called passive immunity, which is not permanent.

Chapter Fourteen: Your Kidneys and the Formation of Urine

There are many parts of your urinary tract, with the main job of this system being to make urine and allow it to pass out of the body. The main organ in this system is the kidney. Your two kidneys filter the blood on a continual basis and reabsorb much of the filtered water and electrolytes. They also allow for the excretion of metabolic wastes, certain drug metabolites, and other substances your body does not need or want.

Your kidneys have other jobs as well that you might not know much about. They're intricately involved with maintaining a normal blood pressure through a group of hormones that also help regulate the amount of water in the body. Kidneys make a hormone called *erythropoietin* that helps promote red blood cell synthesis in the bone marrow. It also promotes the formation of active vitamin D from its less active form.

In this chapter, we will talk about the basic anatomy of the urinary tract, from the kidneys to the urethra where urine leaves the body.

The Anatomy of the Urinary System

The kidneys are bean-shaped organs located on either side of the spine, in the back of the body beneath the rib cage. The kidneys' main job is to remove a molecule called *urea*, which is the major breakdown product of protein metabolism. Each kidney contains millions of small filtering units, known as *nephrons*. The nephron is the single filtering unit of the kidney.

Each nephron consists of a ball formed by small capillaries that go in and out of the area. The ball is known as a *glomerulus*; its sole job is to filter the blood on a 24/7 basis. About 150 quarts of blood get filtered through your kidneys every day. We will talk more about how the filtered blood gets treated within the nephron in a little bit. In the end the urea, together with water, electrolytes and other waste substances, becomes urine that passes into a structure in the kidneys called the *renal pelvis* before exiting the kidneys through the tubes called *ureters*. This is what the kidney structure looks like:

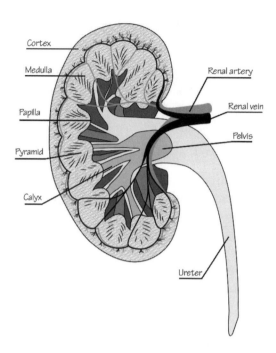

The kidneys are each about the size of a human fist. Only about two quarts of filtered blood end up as urine. This means that a great deal of filtered blood needs to be reabsorbed by the nephron so that the urine is able to be concentrated. The kidneys in the body look like this (notice that the right kidney is below the left because the liver is in the way on the right side):

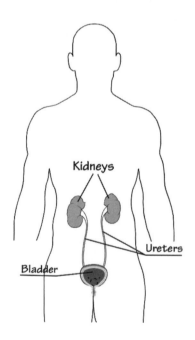

The waste products in your blood come from the normal breakdown of active muscle tissue as well as from the protein in the food you eat. Urea and ammonia are both breakdown products of protein metabolism, and the kidneys are the main option for getting rid of these metabolic waste products.

The kidneys themselves are protected by a fatty layer called *perirenal fat*. The actual outer layer of the kidneys themselves is called the *cortex*. Deep within this is the *adrenal medulla*, which is where the nephrons are located. As the urine gets filtered and processed, it exits out of renal collecting ducts that come together into a *calyx*, where the urine enters the renal pelvis. Each kidney has about 7 to 13 calyces.

The Ureters

The ureters are tiny tubes consisting of smooth muscle fibers that send urine from the kidneys into the bladder. In an adult, the ureters are generally 25-30 centimeters (10-12 inches) long. It is only about 3-4 millimeters in diameter. The ureters begin at the pelvis of each kidney and travel down the psoas muscle to reach the brim of the pelvis. Then they cross in front of the common iliac arteries and enter the back and side portions of the bladder. The openings of the ureters in the bladder are just slits that lead into the *trigone* of the bladder. They are about an inch apart when the bladder is empty and 2 inches apart when the bladder is full. This image shows the ureters as they enter the bladder:

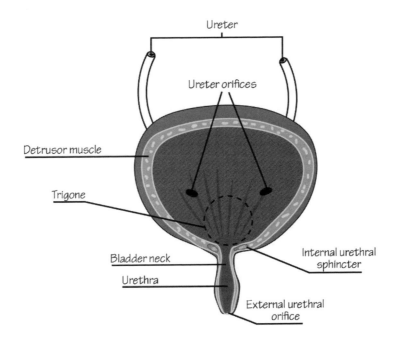

The Bladder

The bladder is a small structure made mainly of smooth muscle that is located in the pelvis, just above and behind the pubic bone. When it's empty, the bladder is about the size and shape of an apple or pear. It holds the urine your kidneys make and send through the ureters. The bladder is the main storage vessel for urine, allowing urination to be voluntary and usually only when it is convenient to do so.

The bladder contains smooth muscle cells that stretch to allow urine to accumulate. These muscle cells then contract in order to empty the bladder. The usual capacity of your bladder is about 400-600 milliliters. There are two sphincters in the bottom of the bladder that allow urine to flow outward. One sphincter is involuntary, while the other one you have control over. Urine exits the bladder into the urethra, which carries the urine to the outside of the body.

The Urethra

The urethra is a tubular structure that connects the bladder and the reproductive organs to the outside of the body. Both semen and urine pass through the urethra in men, who have a much longer urethra. The walls of the urethra contain three distinct layers that are continuous with the bladder.

The main difference between the male and female urinary tract is the length of the urethra. In a woman, the urethra is about 1.5 to 2 inches long. It exits a space between the *clitoris* and the vagina. In men, the urethra runs through the penis and is about 8 inches long. It opens out at the end of the penis.

What Happens in the Nephron?

The nephron is the structural and functional unit of the kidney. It takes millions of these nephrons to do the job of the kidneys as a whole. They are complex and interesting structures that use a creative strategy in order to filter blood, but then reabsorb most of the stuff you need while excreting the things you don't into the urine.

The filtering unit of the nephron is called the *renal corpuscle*. It is a globular structure made from the glomerulus and the *Bowman's capsule*. The structure looks like this:

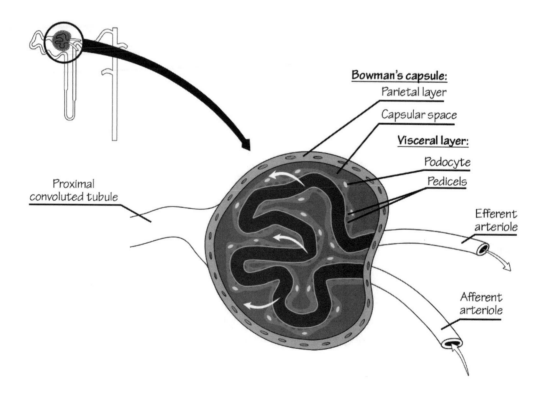

Each renal corpuscle has an outer Bowman's capsule and an inner glomerulus. There is an afferent, or incoming, arteriole which brings the blood into the glomerulus. This incoming blood is met with a very leaky glomerulus that allows small molecules and water to slip through the side-walls and into the nephron, where this fluid is called *glomerular filtrate*.

The glomerulus is designed to be leaky enough for all but proteins and very large molecules to slip through. This means that the glomerular filtrate contains things like glucose, salts, urea, amino acids and other small molecules in a dilute solution of water. Many of these, especially the glucose, salts and amino acids, are things you really need, so your kidneys must have a way to recollect these things further down the line. All blood that isn't filtered will leave the nephron through the efferent, or outgoing, arteriole.

The *glomerular filtration rate*, or *GFR*, is a measure of how well the kidneys are filtering blood. The filtering process depends on the actual pressure of the blood in the arterioles, which is greater than the nephron pressure. There is a natural push to filter blood at that point.

If your blood pressure is too low for any reason, or if there is some urinary blockage that increases the nephron pressure, the kidneys cannot filter and the GFR will decrease. If this is really severe, you might have temporary or permanent kidney failure, depending on the circumstances.

Fun Factoid: *The GFR is the best measure of how your kidneys are functioning. A normal GFR is about 90 millimeters per minute, and renal insufficiency is usually defined as having a GFR of less than 60. Fortunately, the kidneys are still fairly efficient, so you likely wouldn't need kidney dialysis unless your GFR was less than 15 millimeters per minute, which is called* end-stage renal disease.

Now that the blood is filtered, you have a lot of glomerular filtrate that enters the *proximal convoluted tubule* (PCT) of the nephron. This is the start of a journey that involves reabsorbing a lot of salts, glucose and amino acids, as well as much of the filtered water. This image shows you what the process looks like:

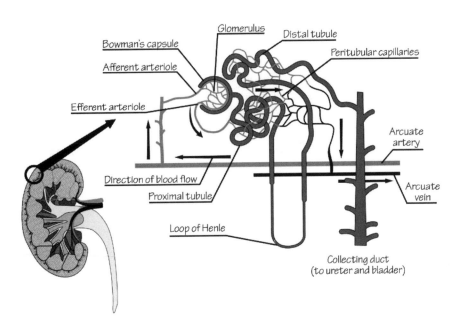

From the PCT, the filtrate extends down into the *renal medulla*, where it goes down the *loop of Henle*. Through a complex process, electrolytes and water get taken back up in these areas so you don't lose them in the urine. Then there is a *DCT*, or *distal convoluted tubule*, that leads further into the collecting ducts. The collecting ducts are the last point where the kidneys adjust the amount of salt and water in the urine before it is excreted.

There are three stages involved in the making of urine by the nephron. These include the following steps, which happen in sequence:

- **Filtration:** This is the first stage of nephron functioning. Your blood pressure forces plasma (the liquid part of blood) through tiny capillaries inside the glomerulus. Blood cells and proteins are too big to pass through the capillary wall, so they stay in the bloodstream. The filtered fluid is then called *glomerular filtrate*. It collects in the capsule and enters the proximal convoluted tubule or just the renal tubule.
- **Reabsorption:** During this stage, the necessary substances your body can't afford to lose in the filtrate reenter the bloodstream again. This process happens in the different renal tubules (the proximal convoluted tubule, the loop of Henle, and the distal convoluted tubule). In the tubules, there are both active and passive processes that allow things like sugar and other nutrients, electrolytes, and excess water to pass back into the surrounding capillaries. Normally, around 100 percent of glucose is reabsorbed back in the bloodstream. If you have diabetes mellitus, however, not all of the glucose can be easily reabsorbed, so some of the glucose spills into the urine. Some reabsorption involves active transport (which needs ATP to function) and other processes involve only passive diffusion.
- **Tubular secretion:** This is the passage of substances out of the capillaries and into the renal tubules and collecting ducts. Tubular secretion is one way that waste materials can enter the urine. Drug metabolites can be secreted directly into the tubules or collecting ducts from the bloodstream. Urea and uric acid that may have been incidentally reabsorbed early on in the process are later secreted again in the distal tubules. Excess potassium ions are also secreted by the distal renal tubules in order to enter the urine. The secretion of acid or bicarbonate into the tubules also helps maintain the pH of the blood.

The amount of urine made varies according to how much water your body needs. *Antidiuretic hormone*, or *ADH*, from the posterior pituitary gland controls the amount of urine made. If you sweat a lot or don't drink enough water, there are *osmoreceptors* in your hypothalamus that sense you don't have enough water in your bloodstream. The osmoreceptors then cause neurosecretory cells in the hypothalamus to make ADH. This is the hormone that tells the kidneys to hang onto more water. This is regulated in the collecting ducts of the kidneys and, if ADH levels are high, will decrease the amount of urine you make. If you instead take in too much water, osmoreceptors detect that you are overhydrated and will lower the ADH level to make more urine.

Hormones in the Kidneys

The kidneys have other important jobs to do besides making urine. They make hormones involved in a variety of different bodily functions. While the kidneys are mainly made from nephrons, there is other kidney tissue called *parenchymal tissue* that have non-urine-making functions. Here's what else the kidneys make:

- **Calcitriol:** Calcitriol is the active form of the vitamin D (called *D3*) in humans. It is made in the kidneys using the precursor molecules of vitamin D you get when UV light from the sun strikes your skin. Calcitriol works alongside parathyroid hormone to increase calcium ions in the blood.

When the concentration of calcium ions drops too low, the parathyroid glands secrete parathyroid hormone. Calcitriol functions to promote absorption of calcium via the small intestines. It also tells the osteoclasts of the skeletal system to break down bone to release calcium ions in the blood.

- **Erythropoietin:** This is also known as *EPO*. It's a hormone made by the kidneys to stimulate the production of red blood cells by the bone marrow. The kidneys can monitor the amount of red blood cells in the blood that passes through the capillaries. When the blood becomes hypoxic (low in oxygen) or the kidneys sense that there are too few erythrocytes in the blood, the synthesis of erythropoietin initiates RBC production. People with kidney failure often have anemia because there isn't enough erythropoietin being made by these damaged kidneys.
- **Renin:** Renin is not actually a hormone but is instead an enzyme the kidneys make to start the *renin-angiotensin-aldosterone system*. This system increases blood volume and blood pressure in response to a perceived decrease in blood pressure, dehydration or blood loss—all of which are things the kidneys can detect. Under these conditions, the renin is sent into the bloodstream, where it turns *angiotensinogen* from the liver into *angiotensin I*. This is further turned into *angiotensin II* in the lungs, which stimulates the adrenal cortex to make aldosterone and increase the reabsorption of sodium ions and water into the bloodstream. This is what it looks like:

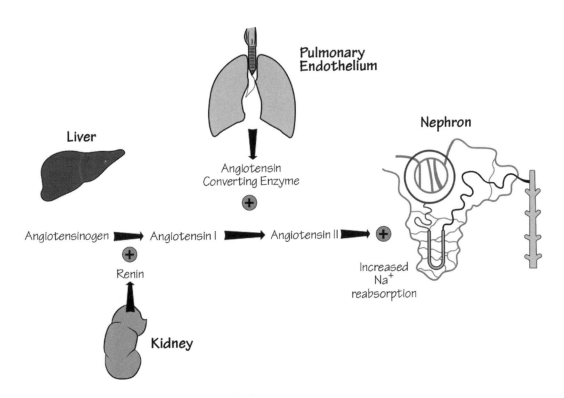

To Sum Things Up

The urinary or renal system is the main way many metabolic wastes are filtered out of the bloodstream. Many of these metabolic wastes come from the breakdown of protein metabolism and involve both urea and ammonia. The kidneys also secrete either acidic or alkaline ions in order to regulate the acid-base levels in the bloodstream in order to keep the pH normal. This process works along with the lungs, but is a much slower process on this end.

The nephron is the functional unit of the kidney's filtration system. There are millions of these. They consist of glomeruli that filter the blood continually. The glomerular filtrate that comes from this process gets further modified in the different tubules of the nephron in order to resorb most of the water, almost all of the amino acids, and glucose back into circulation, while keeping the waste products in the urine. Near the end of the process, water and sodium levels are fine-tuned under the influence of antidiuretic hormone, so that just enough water is peed out in the urine without impacting your blood volume and pressure.

The kidneys participate in making erythropoietin to turn on red blood cell production. They're also actively involved in turning precursor vitamin D into vitamin D3, which is the active form of the vitamin. Finally, the kidneys are intricately involved in controlling blood pressure by making the renin enzyme.

Chapter Fifteen: The Reproductive System

The reproductive system is yet another example of coordination in the amazing human body. In this case, the coordination must happen between the male and female reproductive systems just as much as it needs to happen within every man and woman by themselves.

The Female Reproductive System

The main organs of the female reproductive system are the ovaries, the *Fallopian tubes*, the uterus, the vagina, the *vulva* and the *mammary glands*. Most of these structures are directly involved in the making and transportation of the female egg from the ovaries to the point of fertilization in the Fallopian tubes. The ovaries are the female gonads that make the egg cell, or female gamete, which is destined in some cases to be fertilized by the male gamete, or sperm cell, if conditions are right. These are the basic anatomical structures of the female reproductive tract:

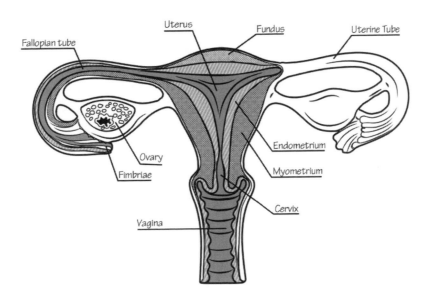

The ovaries are also responsible for the production of sex hormones; however, none of this would happen without the influence of the hypothalamus and the pituitary gland on the ovarian tissues. The female reproductive system is also different from the male reproductive system because this is where the egg is fertilized, and the resulting zygote is allowed to grow and develop first into an embryo and then into a fetus. At the end of many months in the uterus, the fetus emerges in the birth process to become a baby that is nourished by breastmilk.

The ovaries are a pair of glands located on either side of the uterus. They are about the size and shape of almonds. They connect to the uterus via the Fallopian tubes but are also heavily supported by ligaments that keep them in place. There are actually several ligaments, including the *broad ligament*, that hold the whole female reproductive tract in its proper place in the pelvis.

The main job of the ovaries is to make egg cells, or *oocytes*, which are meant to be fertilized by sperm. In order to do this, they must also make reproductive hormones such as estrogens and progesterone, which help in this process and contribute to the secondary female sex characteristics you can see, like breasts, female fat distribution, and female genital structures.

Every month, at the time of ovulation, the ovaries release at least one mature egg. The egg travels from the ovary into the Fallopian tube, where it may or may not be fertilized by a male sperm. If it is not fertilized, it degenerates and menstrual flow occurs, completing the menstrual cycle. In women, the hormone levels of FSH, LH, estrogen and progesterone vary throughout the menstrual cycle, in order to create a true cycle.

The Fallopian tubes are a pair of muscular tubes that come out of the top sides of the uterus and extend outward to reach the ovaries. Each of these end in a funnel-shaped structure known as the *infundibulum*, which has many finger-like projections (called the *fimbriae*). The fimbriae sweep over the outside of the ovaries at the time of ovulation, allowing the egg to enter the infundibulum and pass down the tube. The inside of the Fallopian tube is covered in tiny cilia that work alongside smooth muscle to carry the egg from the ovary to the uterus.

The uterus is a pear-shaped, hollow, muscular organ located behind and above the urinary bladder. It is connected to the Fallopian tubes at its upper end and to the vagina at its lower end. The cervix is considered part of the uterus, and sits in the opening through which the contents of the uterus pass in order to be expelled by the vagina. The inside lining of the uterus is known as the *endometrium*. It provides support to the embryo during gestation and changes in thickness throughout the menstrual cycle. The visceral or smooth muscles of the uterus contract during childbirth to help expel the infant in the birthing process.

The vagina is a muscular, elastic tube that connects the cervix of the uterus to the outside of the body. It is located beneath the uterus and behind the urinary bladder. The vagina acts as the receptacle for the penis during intercourse and carries the sperm to the Fallopian tubes and uterus. It is also the birth canal for an infant, and it therefore needs to be stretchy and elastic. It is also the passageway for menstrual blood during menses.

The vulva is what we call the *external female genitalia*. The vulva surrounds the external aspect of the opening to the urethra and vagina. It includes the clitoris, the *labia minora*, the *labia majora* and the *mons pubis*. The mons pubis is a raised layer of fatty tissue between the pubic bone and the skin. It provides cushioning for the vulva. The lower part of the mons pubis divides into a right or left half, each of which is a labium majora. The mons pubis and the labia major contain pubic hair. Inside the labia majora are smaller, hairless folds of skin known as the labia minora. They surround the urethral and vaginal openings. On the upper end of the labia minora is a small mound of erectile tissue called the clitoris, which is the source of a female's sexual pleasure.

The female breasts are mostly fatty structures on the chest, but they also have mammary glands that are used for the production of milk. There are two breasts, one located on either side of the thoracic region at the front of the body. In the middle of each breast is a pigmented nipple that sends milk from the mammary glands to the outside when stimulated by infant suckling. The areola is a thickened, pigmented band that surrounds the nipple and protects the tissue beneath it.

The mammary glands of the breasts are a specialized type of gland that is used to make milk to feed infants. Inside each breast are 15-20 clusters of mammary glands that become activated during pregnancy and remain active until milk is no longer necessary. It takes prolactin to make milk and oxytocin to allow the milk to flow during the act of infant suckling. This is what the breast looks like:

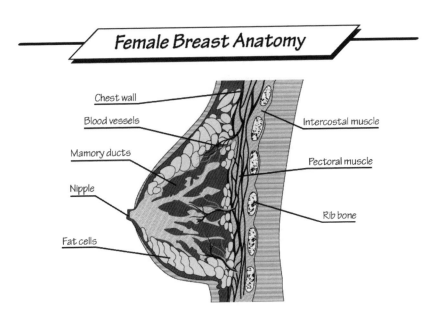

The Menstrual Cycle and Pregnancy

While the female reproductive system seems busy and complicated, it really has only two physiological goals: to create both an oocyte and a fertile environment for the zygote that arises after fertilization. This zygote can then grow and develop in the uterus. The whole menstrual cycle is about preparing for this pregnancy, even though pregnancy likely only happens to a woman at most a few times in her life.

If an egg is produced but isn't fertilized, the lining of the uterus is shed in the form of menstruation. The act of menstruating is basically a resetting of the menstrual cycle. The entire reproductive cycle takes about 28 days, but may be as short as 24 days or as long as 36 days, depending on the woman.

Ovulation and egg formation involves follicle stimulating hormone (FSH) and luteinizing hormone (LH). The ovaries ovulate or release a mature egg or ovum that has been developing for weeks prior to its ultimate release. Many follicles in the ovary develop at the same time, but only one reaches full maturity (while the others will involute and disintegrate). By about 14 days into the menstrual cycle, the egg achieves the necessary maturity level and is released from the ovary at the time of ovulation. This is what it looks like:

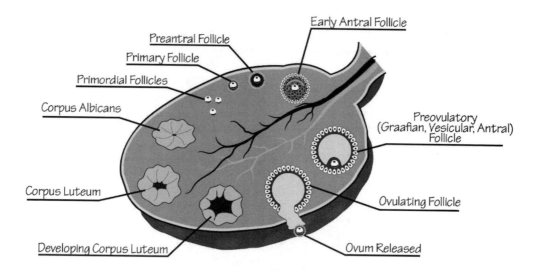

As you can see, the follicle is where the egg cell comes from. After the egg is released, the follicle turns into the corpus luteum, which makes progesterone to support the pregnancy until the *placenta* can make this hormone. As the corpus luteum fails, it becomes white in color and is called the *corpus albicans*, which is not really an active hormone-making part of the ovary.

Fertilization is the process of combining the sperm and egg to make a zygote. The egg is released from the ovary and is swept into the fimbriae, where it catches the egg and sends it down the Fallopian tube to be fertilized. Most fertilizations happen in the midportion of the Fallopian tube. It takes one week for the fertilized egg to reach the uterus. If the egg is fertilized by sperm, it becomes a zygote and finally an early embryo that implants to the uterine wall, resulting in a pregnancy.

Fun Factoid: One common but unfortunate cause of infertility is called a luteal phase defect. *The corpus luteum absolutely must make enough progesterone for a minimum of twelve days so that the implanted embryo can start to make its own pregnancy-supporting hormones. Some women commonly have their corpus luteum give out prematurely so that the pregnancy is lost before it can even take hold. This is sometimes treated by giving the woman extra progesterone in order to make up for the missing hormone later into the menstrual cycle.*

Pregnancy happens when a fertilized egg becomes a zygote, an embryo, and finally a fetus. It is the *blastocyst*, which is an early embryonic form of the growing baby, that implants into the uterine wall and grows for about 9 months. When the blastocyst implants into the uterine lining, it begins to form a placenta, an *umbilical cord*, and an *amniotic cavity*. For the first eight weeks, the embryo develops all the structures it needs to become a human, while the rest of the pregnancy involves maturing and growing these organs. The fetal organs become more complex until the baby is ready to be born during the childbirth process.

Lactation, or *breastfeeding*, involves the production and release of milk to feed a baby. The production of milk begins before birth and is governed by the hormone known as prolactin. Prolactin is produced in response to an infant sucking on the nipple. Milk is made as long as there is active sucking on the nipple. When the infant is weaned off breastfeeding, prolactin levels decrease and milk production stops. The release of milk by the nipples is also called the *milk letdown reflex*. It is controlled by the female hormone oxytocin. Oxytocin is released in response to the suckling of an infant so that milk is only produced when it is needed.

Hormones and the Ovaries

The ovaries of mature females secrete a mixture of estrogens and progesterone, making them basically endocrine glands that interact with other endocrine glands. The most prevalent estrogen type in fertile women is called *17-beta-estradiol*. Estrogens are considered steroids that are responsible for the conversion of young girls into mature women.

Estrogens help to make the female secondary sexual characteristics, such as the increase of fatty tissue in the female tissues, the growth of pubic and axillary hair, the broadening of the pelvis, the further development of the uterus and the vagina, and the development of breasts. They also help increase the thickness of the uterine lining in preparation for a pregnancy, should one occur.

Estrogen also has non-reproductive functions. It blocks the effect of parathyroid hormone, minimizing the loss of calcium from bones and thus helping the bones to remain strong. Estrogens also promote the clotting of blood. This can sometimes be a bad thing, because high-dose birth control pills with a lot of estrogen in them increase the risk of clotting events, such as a stroke or pulmonary embolism.

Progesterone is also a steroid. It has numerous effects in the body, some of which are related to reproduction. It serves to mature the uterine lining in the latter half of the menstrual cycle, and remains elevated throughout pregnancy.

Steroids like progesterone and estrogen are hydrophobic molecules that cannot be transported in the aqueous blood alone. They must be bound to a serum globulin in order to enter the bloodstream. In target cells that have receptors for these hormones, the estrogen or progesterone bind to receptors inside the cell rather than on the cell surface. These steroid hormones enter the cell nucleus and bind to DNA in order to change which proteins the cell makes.

The synthesis and release of estrogens is stimulated by FSH, which is itself turned on by GnRH released by the hypothalamus. Luteinizing hormone is secreted by the anterior pituitary gland at the time of ovulation. This hormone also depends on GnRH secretion by the hypothalamus. There is an LH surge at the time of ovulation that triggers the release of a mature egg.

Fertilization of the egg also occurs under the influence of progesterone. Sperm cells swim by means of *chemotaxis*. This is basically the act of following a chemical signal in the body. The sperm cells travel to areas where there are high levels of progesterone secreted by cells surrounding the egg. The progesterone opens channels in the plasma membrane of the sperm cell, causing the sperm's *flagellum* to beat more rapidly and vigorously toward the egg.

Fertilization involves one sperm cell entering the thick *zona pellucidum* layer around the egg cell. When this happens, a reaction called *the acrosome reaction* occurs. This effectively blocks any more sperm from entering the egg. This is necessary because, should more than one sperm cell enter the egg, it would not create a viable zygote or fertilized egg. This is what fertilization looks like:

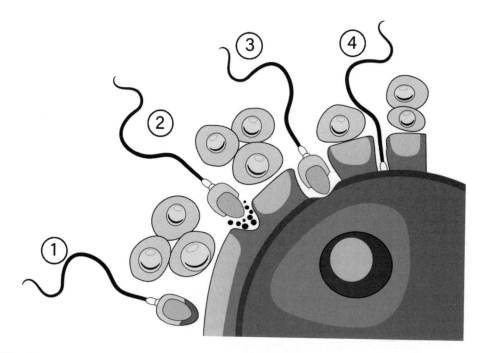

When the fertilized egg passes down the fallopian tube, it undergoes its first mitotic divisions to create a zygote. By the end of the first week, the developing embryo becomes a hollow ball of cells that are known as a *blastocyst*. At this time, the blastocyst enters the uterus and embeds into the endometrium in a process known as *implantation*. After implantation, the pregnancy begins. This is what a blastocyst looks like as it develops from a zygote:

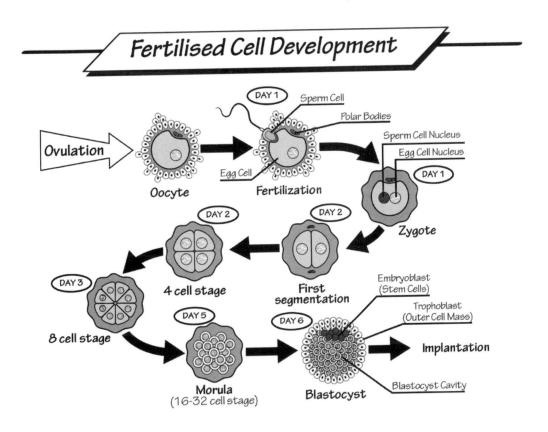

Fertilised Cell Development

The blastocyst consists of two parts: the inner cell mass that forms the fetus, and the *trophoblast* that ultimately becomes the umbilical cord. The trophoblast begins to secrete human chorionic gonadotropin. HCG is a glycoprotein similar to FSH and LH, but is not blocked by the rising progesterone levels in pregnancy. It prevents the deterioration of the corpus luteum, which continues to make progesterone to support the pregnancy until the placenta takes over production.

Because the cells that will become the placenta produce HCG, its appearance in the urine makes for a good early test to confirm a pregnancy. It can be detected in the blood prior to a missed period. Other hormones of the female reproductive system include *relaxin*, which is the hormone that relaxes the *symphysis pubis ligaments* and helps to enlarge and soften the opening of the cervix during the later stages of pregnancy. It's at its highest at the end of the pregnancy. It also seems to promote new blood vessel formation, and probably plays an important role in the interface or connection between the placenta and uterus early in the pregnancy.

The Male Reproductive System

The *male reproductive system* is simpler than the female reproductive system, mainly because there's no cyclic aspect to it. Spermatogenesis, or the formation of sperm, happens every day, with about 200 to 300 million spermatozoa created daily. It takes about 74 days for a *spermatogonium*, or *primordial sperm cell*, to differentiate into a gamete, and about 3 months in total for the primordial cell to fully develop and go through the male ducts to the outside at the time of ejaculation.

Male Sexual Anatomy

The male reproductive system consists of the testes, the *epididymis*, the *vas deferens*, the *seminal vesicles*, the *prostate gland*, and the penis/urethra. The Leydig cells in the testes make testosterone, which is necessary for the development and maturation of male sexual attributes, and contributes to sperm production. The sperm are first produced in the testes and then matured inside the epididymis. The sperm then travel out of the epididymis via the vas deferens. This is the overall structure of the male reproductive tract:

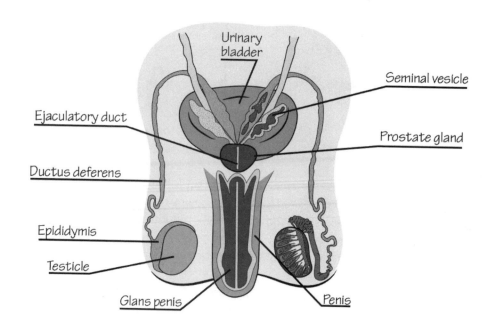

Spermatogenesis

Spermatogenesis is the process in which sperm cells start out as immature stem cells called *spermatogonia* and become mature spermatozoa. We talked earlier about meiosis, which is how a diploid cell becomes a haploid gamete. This is what happens in the *seminiferous tubules* of the testes:

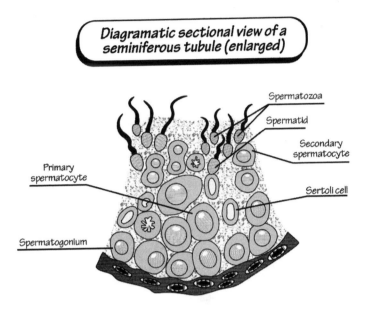

The first cells in the pathway are spermatogonia, which create two separate *spermatocytes* through the process of mitosis. The primary spermatocyte divides into two different secondary spermatocytes by undergoing meiosis I. These then develop into *spermatids*, which do not become mature spermatozoa until they develop a tail. They are not considered motile sperm until they mature further in the epididymis, which is the next stop past the seminiferous tubules. The pathway that sperm cells take in the testes is shown here:

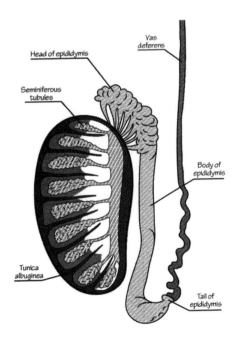

Sperm cells start their development in the seminiferous tubules, travel through the *rete testes* or straight tubes, through the straight tubules, and finally into the epididymis, which is located behind the testes. The epididymis is where the sperm cells mature and gain motility before exiting the vas deferens.

The vas deferens is the tube that travels through the groin. There are two of these that come together near the center of the body, where each one forms an enlargement called the *ampulla* just before reaching the prostate gland. The vas deferens connects to a duct from the seminal vesicle to make an *ejaculatory duct*. There are two of these short ducts that pass through the prostate gland and exit as a single urethra.

The urethra is the exit tube for both urine and semen in males. It extends from the base of the bladder and has three separate segments. The proximal part is called the *prostatic urethra*, which goes through the prostate gland. The ejaculatory duct and spermatozoa join in and then exit to make the *membranous urethra* that goes through the pelvic floor. Finally there is the long *penile urethra*, or *spongy urethra*, which passes through the penis itself and finally to the outside of the body.

Sperm cells that leave the epididymis are mature, but are not the same thing as the *semen* that is ejaculated at the time of intercourse. There is a lot more to semen than just sperm cells. This is what it takes for these cells to become what is properly called semen:

- In the *Sertoli cells*, or *nurse cells*, of the seminiferous tubules, the cells will secrete a fluid that helps to provide a nourishing liquid for spermatozoa to travel to the epididymis and through the vas deferens. This fluid is only about 2 to 5 percent of all of the semen volume.
- Next comes the seminal vesicles, which provide up to 75 percent of the seminal fluid. This is where fructose sugar is added to semen, which is what provides a type of carbohydrate nourishment for sperm cells until they reach their destination. Citrate, some enzymes, and amino acids also get added, along with molecules that help reduce a woman's immune response against the foreign sperm cells.
- The prostate gland then adds about 25 to 30 percent of the fluid that becomes semen. This adds some acidity to the semen as well as zinc, which is said to stabilize the DNA inside the sperm cells. This is why zinc is important for healthy sperm. The prostate gland also adds *acid phosphatase*, an essential enzyme.
- Lastly, there are tiny *bulbourethral glands* that contribute just 1 percent or less of semen volume. They add mucus that lubricates the urethra so that sperm cells pass more easily. The mucus also creates a plug of semen in the vagina to prevent sperm cells from falling out after being deposited. After a short time, however, this mucus plug breaks up and sperm can travel up through the vagina, cervix and uterus until they reach the Fallopian tubes, where fertilization takes place.

The normal male sperm should be about 2 milliliters of total volume per ejaculate (or greater). The pH range is about 7.2 to 8.0 with about 20 million sperm cells per milliliter or more. Lab technicians look at sperm motility and the sperm *morphology* (the shape of sperm). Ideally, at least 50 percent of sperm should be able to move forward, and there shouldn't be too many spermatozoa that are abnormally shaped. A normal sperm cell looks like this:

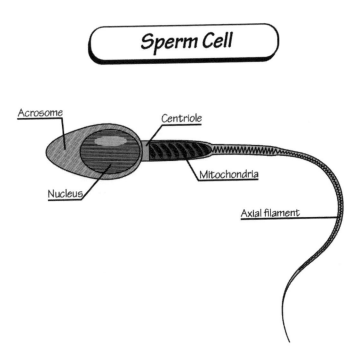

As you can see, there isn't much cytoplasm in a sperm cell. The main cell body has a nucleus but not much else. In front of the cell body is an acrosome with enzymes which will facilitate the ability of the sperm to get through the *zona pellucida* surrounding the egg cell. There is a centriole that serves as the motor driving the *axial filament*, or "tail" of the sperm. In the space between the head and the tail are many mitochondria that provide the necessary energy to drive the tail. None of these mitochondria enter the egg cell; so during fertilization, all the mitochondria you receive genetically come from the egg.

To Sum Things Up

The reproductive system is a complex network of structures that actually starts in your brain's hypothalamus. This is where the first hormone, GnRH, is released, which triggers the pituitary reproductive hormones, including FSH and LH. Both act in males and females but have different jobs in each gender.

The female reproductive system is cyclical in its physiology and is based on the monthly menstrual cycle. The ovaries make about one mature egg cell per month. It's released from the ovary due to the influence of LH, travels to the Fallopian tube, and gets fertilized in some cases with a male sperm cell. If the egg isn't fertilized, it disintegrates; and the uterine lining, which has built up throughout the month, eventually sheds during the menstrual period.

If the egg cell is fertilized, the zygote made in the process divides into several cell stages to become a blastocyst. The blastocyst embeds into the uterine lining which has prepared for it during the month, where the pregnancy begins and the embryo continues to grow until the time of birth.

In men, sperm cells are created in the testes within structures called seminiferous tubules. The sperm cells created there will become motile in the epididymis and travel through the vas deferens. Along its travels, the sperm cells become semen, or seminal fluid, by getting nutrients and other substances that help the sperm effectively exit the urethra during ejaculation and travel through the female reproductive tract.

In that sense, it takes a coordinated female reproductive system and heathy sperm to come together at the right time in the menstrual cycle in order to create new life. The timeframe for fertility in a given menstrual cycle is just a few days, so it really takes a bit of luck and knowledge about when a pregnancy is most likely to occur in order to get pregnant or avoid so, depending on one's wishes.

Chapter Sixteen: The Basics of Your Skin

Your skin probably doesn't seem terribly important in the span of things. Yet, there are many who consider the skin to be the largest organ of the body. About 12 to 15 percent of your total body weight is made from skin, and it covers basically everything within your body, providing vast amounts of protection you couldn't live without.

The skin system is also called the *integumentary system*. Besides the skin itself, the hair and nails are a part of this system. Overall, the skin does far more than just cover your body. It's made of keratinized epithelial cells that help to waterproof your body. It cushions the deeper structures beneath it and allows you to sweat, which excretes waste products and regulates your body temperature. Hair follicles do more than just look pretty by making hair. The skin itself has many nerve receptors that provide all kinds of sensation to the body. Vitamin D precursor molecules are in the skin and start the process of making active vitamin D for many body systems.

The Anatomy of Skin

Your skin looks different depending on the part of the body. You have what's called *glabrous skin*, which is hairless. Other skin areas will have a thick epidermal layer with ridges, such as those on the pads of your fingers or toes. The skin on your face and the back of your hands is much thinner than the skin on the palm of your hand or the soles of your feet. Areas with a lot of potential for friction will be thicker than areas where friction is less likely.

The outer layer of skin is called the *epidermis*. This is mostly made of dead epithelial cells and has no capillaries or nerve endings in it. The epithelium is called *stratified squamous epithelium*. The term *stratified* means it's layered, while the term *squamous* means that the cells are flattened in shape. There are several layers of epithelium that start at the base and extend outward to an outermost layer of dead, keratinized cells that slough off on a regular basis. The term *keratinized* means there's a lot of the tough keratin protein in this skin layer that will protect the cells from damage.

This is an image of your epithelium and its layers:

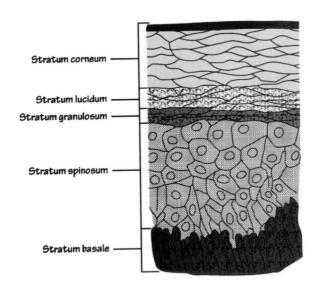

As you can see, the innermost layer is the *stratum basale*. These are mitotically active cells that give rise to new cells that get pushed up into the *stratum spinosum*, which contains spiny cells. Above this is the *stratum granulosa* layer, which begins to be keratinized. As the cells get pushed out further, there is the *stratum lucidum*, a thin layer that also contains a precursor protein to keratin. Finally, there is the *stratum corneum*, which has few nuclei, is highly keratinized, and is mostly waterproof.

The stratum basale has a protein-containing basement membrane beneath it that marks the boundary between the epidermis and the *dermis* beneath it. The pigmented cells, or melanocytes, are located in this basal layer. The dermis is highly vascularized and contains all of the main structures you associate with skin, such as the nerve receptors, hair follicles, sweat glands and oil glands. This layer covers the *hypodermis*, a deep, fatty layer made of connective tissue and fat cells. The hypodermis is mostly important in protecting underlying structures, because it has a fatty cushion.

This image shows the structures of the dermis:

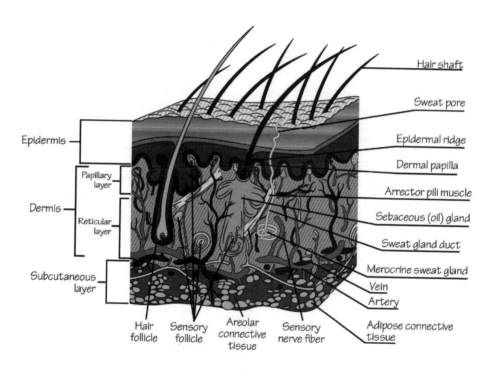

The different structures include:

- **Basket cells:** These cells sense pressure at the base of each hair follicle.
- **Arteries and veins:** These are the blood vessels that supply each cell of the dermis. As mentioned, there are no blood vessels in the epidermis, which must instead rely on the diffusion of nutrients over the basement membrane for nutrition.
- **Hair follicle:** This is a sheath with a bulb at the base where hair originates and is nourished. The hair follicle extends up into the epidermis as well, and leads to the hair shaft.
- **Erector pili muscle:** This is the tiny muscle attached to each hair follicle that contracts to form goosebumps when you're scared or cold. The idea is to better diffuse sweat and get rid of body temperature faster when these muscles are contracted.

- **Langerhans cells:** These are immune cells in the skin that attach to antigens in any damaged skin, alerting the rest of the immune system that a potential infection is present.
- **Merkel discs:** These are sensory receptor organs that detect light touch sensation.
- **Pacinian corpuscles:** These are also sensory receptors that are located in skin, that respond mostly to vibration but also to pressure applied to the skin.
- **Sebaceous glands:** These are sac-shaped glands that release the oily substance found on much of your skin. The oil will protect the hair shaft so the hair won't be so brittle.
- **Sweat glands:** These are also called *sudoriferous glands*. They produce two kinds of sweat. There is the sweat most of your body releases in order to regulate your body temperature, and there is the sweat in your axilla and other body areas that have more odor to the secretions.

Summary of Skin Functions

Your skin has a lot of simultaneous jobs to do as part of protecting your body and regulating its temperature. It is also a huge sensory organ that allows you to have a great awareness of your environment. People born with skin insensitivity to things like pain and temperature have a high rate of injury and death, because they cannot tell if the skin is in danger or not.

These are the main functions of your skin:

- Protect the body from UV radiation
- Form new cells in the repair of minor injuries
- Help maintain the body's shape
- Help store vitamin D, glucose, fat and water
- Generate vitamin D through ultraviolet light exposure
- Secrete melanin to protect the body from sunburn
- Act as a receptor for touch, pressure, pain, heat and cold
- Help secrete waste materials through perspiration
- Protect the body from changes in temperature
- Protect the body from dehydration
- Protect the body against infectious diseases
- Protect the body's internal living tissues and organ systems

Temperature Regulation

Your skin plays a major role in regulating your body temperature throughout a wide range of environmental temperature variations called *thermoregulation*. Constricted blood vessels will retain heat, while dilated blood vessels allow for loss of heat. When *arrector pili* muscles contract, they cause the hairs of the skin to stand up. This causes greater trapping of heat into the layer just above the hairy skin. In addition, vasoconstriction of the vessels in the skin can be used to keep heat inside the body and maintain homeostasis. The body can remain warm and at about the same temperature regardless of the external temperature, especially if you also shiver to retain heat.

This image shows the processes of regulating body heat in higher environmental temperatures:

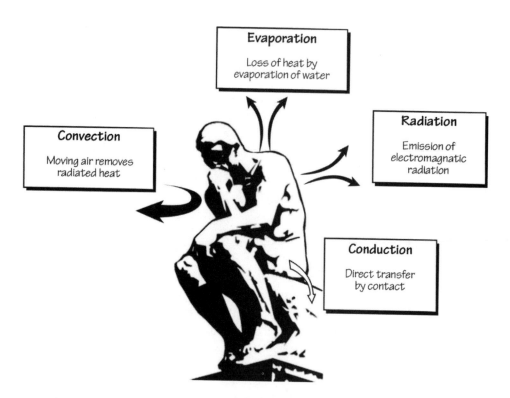

When you are too hot, water can cross outward through the skin by means of the sweat glands. It then dissipates into the air and cools the body through a process known as *perspiration* or *evaporation*. *Vasodilation* is when your blood vessels dilate and circulation increases. This brings heat near the surface of the skin. The enlarged peripheral vessels allow for more blood to pass near the surface of the skin, cooling off the blood as it passes through.

Perspiration is one of the main ways you regulate your body heat. If the outside temperature is above 37 degrees Celsius, or 98.6 Fahrenheit, you will begin to sweat in order to cool the body down. People sweat from all parts of the body but the main sweat glands are located within the axilla and the groin.

Humidity in the environment plays a role in thermoregulation by limiting the evaporation of sweat and controlling heat loss. When you are in a hot and arid environment (like a desert), the gradient between the humidity on your skin surface and the environment is higher, so you lose heat faster when sweating. However, it's harder to sweat and actually have it evaporate in a humid environment.

Convection can also help to cool the body when you are exposed to water or air. The higher your body surface area, the higher the speed of the circulating air around it. The smaller the distance between the blood vessels and the skin surface area, the greater the loss of body heat through the convection process. This means that adults are better at convection compared to children, who have a smaller skin surface area. *Conduction* happens when you touch a cold object. You lose heat by direct contact with the cold surface.

The processes of convection and radiation work much better when the environmental temperature is below 20 degrees Celsius (68 Fahrenheit), while evaporation works much better when the environmental temperature is above this.

Human Hair

The human hair follicle is a unique structure and there's a lot we don't know about it. There are three regions to the hair follicle: the lower segment, or bulb; the middle segment, or *isthmus*; and the upper segment, also known as the infundibulum. The lower segment extends from the base of the follicle to the insertion of the erector pili, or arrector pili, muscle. The middle segment is a short section that extends from the insertion of the erector pili muscle to the entrance to the sebaceous gland duct. The upper segment goes from the entrance of the sebaceous gland duct to the *follicular orifice*. This is what it looks like:

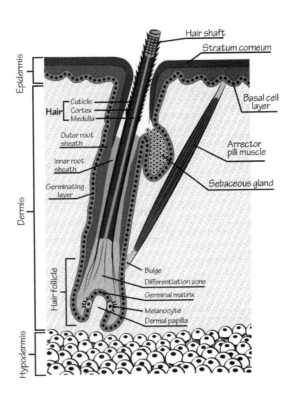

Nails

Your nail is an appendage of the skin. It's the most obvious aspect of the tips of the fingers and toes. It contributes to touch sensation, by acting as a counterforce to the fingertip pad. It is a complex structure with many different parts, as shown here:

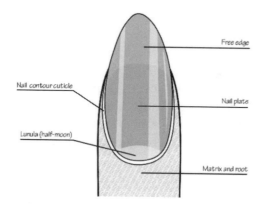

The *nail folds* are the soft tissue aspect on the proximal and lateral aspects of the nail itself. They are not part of the nail but instead your fingers and toes, near the *nail plate*. The most proximal part protects the *nail matrix*, which is where the nail grows from. The *mantle* is the skin that covers this matrix and the proximal part, or base, of the nail plate. The *cuticle*, or *eponychium*, grows outward from the proximal nail bed and sticks to the nail plate. This also protects the nail matrix.

The nail matrix can be seen beneath the *lunula*, which is the crescent-shaped white structure on the proximal nail. This is the part of the nail that makes the keratinized nail plate, and is the part that has melanocytes in it. There are nail-making cells called *onychocytes* that get pushed outward to form the actual nail plate. There is some distal nail matrix but it doesn't contribute much to the making of the nail itself.

The nail plate is what you see as being the nail itself. This is highly keratinized, very hard, and formed from compacted onychocytes arranged in layers, with ridges that are longitudinal. The nail bed and nail matrix are beneath the nail plate itself. The juncture between the nail folds and nail bed has attachment points that prevent the nail itself from loosening easily.

The nail bed itself starts at the distal part of the lunula and ends at the *hyponychium*. It also has ridges like the nail plate above it. The purpose of these is to increase the surface area of the connection between the nail bed and nail plate. The cells of this layer do not make keratin unless your nail falls off for some reason. There is a collagen-containing dermis beneath this layer that attaches it to the bone of the distal phalanx.

To Sum Things Up

Your skin, or integumentary system, is an important part of your body. There are two main layers, the epidermis and dermis, although some people include the hypodermis or subcutaneous tissue as being a third, deeper skin layer. The epidermis is the most protective part of the skin because it's keratinized and waterproof, but it has no circulation of its own.

The dermis has all the typical skin structures, including blood vessels, nerve structures, hair follicles, sebaceous glands, sweat glands, and the arrector pili muscles. These perform many of the skin's basic structures. The skin is a major way of controlling body temperature, and the dermal structures have a great deal to do with this process.

Final Words

Hopefully, you will now feel a great deal more like an expert on your body, including all its major anatomical structures and physiology. While we talked about each of the body systems separately, they are all intricately connected through things like the circulatory system, nerves and endocrine hormones.

The key feature to remember when you think back on what you've learned is that the form of the body is optimized to maximize its function. The bones and muscles are connected in ways that allow maximal movement and minimal stress on your system. There are some anatomical weak points in the body, such as your neck and low back, which are made well enough but are very prone to getting arthritic over time due to everyday wear and tear. Your heart and brain are weak points because their circulation doesn't have a lot of redundancy, making you susceptible to strokes or heart attacks.

Your physiology is extremely well-coordinated so that the body maintains homeostasis and adaptability to your environment. There are feedback loops everywhere in the body, as well as sensors like the osmoreceptors in the hypothalamus that detect how dilute your blood is. Each of these things help to make your body a well-oiled machine that is regulated on many levels to ensure your survival and adaptability.

Should you later decide to study disease states, or *pathophysiology*, you can use what you now know to see how things can go awry if a person gets sick, has a genetic defect, or subscribes to a lifestyle that doesn't support ongoing wellness. Congratulations for making it though this journey of what your body looks like and how it actually works!